Baratz, Morton S.

The union and the coal industry

DATE			
OCT 24'79			
AUG 19'81			
MAY 3 '84			
AUG 07 1998			
MAR 22 2010			
FEB 25 2013			

THE UNION AND THE COAL INDUSTRY

THE UNION AND THE COAL INDUSTRY

BY MORTON S. BARATZ

KENNIKAT PRESS
Port Washington, N. Y./London

THE UNION AND THE COAL INDUSTRY

Reprinted 1973 by Kennikat Press in an unaltered
and unabridged edition with permission
Library of Congress Catalog Card No.: 72-85308
ISBN 0-8046-1721-X

Manufactured by Taylor Publishing Company Dallas, Texas

To C.L.B.

ACKNOWLEDGMENTS

THIS STUDY bears to its great advantage the mark of a number of persons. I regret that my appreciation to them can be shown only by a partial listing of their names.

Within the bituminous coal industry I am deeply indebted to Messrs. G. A. Lamb, Pittsburgh Consolidation Coal Company; G. H. Seal, C. H. Sprague and Son Coal Company; V. M. Johnston, Appalachian Coals, Inc.; M. L. Garvey, Southern Coal Producers Association; J. A. Patterson, Spicer Ice and Coal Company; H. L. Rowe, New London (Conn.) Coal Company; and S. G. Felderman and J. C. Ford, Joy Manufacturing Company.

Invaluable help was given me by the United Mine Workers, particularly by John L. Lewis, President, and John Owens, Secretary-Treasurer. I acknowledge in addition the cooperation of T. W. Hunter of the United States Bureau of Mines.

I have profited greatly from discussion with members of the Yale Department of Economics. Notable contributions to the manuscript came from Professors N. W. Chamberlain, W. Fellner, K. T. Healy, C. E. Lindblom, L. G. Reynolds, E. V. Rostow, and J. Tobin. The greatest single contribution came, however, from Professor J. P. Miller. I am deeply grateful for his patient criticism and counsel as this study progressed from prospectus to book.

I am indebted, too, to R. T. Hannegan, Yale 1953, for his help in checking details, and to the staff of Palmer Library, Connecticut College for Women.

I reserve for the last my gratitude to the Department of Economics for the subsidy which made possible the publication of this book.

MORTON S. BARATZ

Silliman College, Yale University
June 1, 1954

CONTENTS

TABLES

Text

Appendix

CHARTS

INTRODUCTION

ALTHOUGH much has been written about the bituminous coal industry in the past 30 years, few commentators have attempted a systematic analysis of the industry. In particular, although it is universally recognized that the United Mine Workers has dominated affairs in bituminous coal mining for at least the past 20 years, there has been no study of the nature of the union's role, and it is this task to which this work is devoted.

THE PROBLEM

The bituminous coal industry is highly competitive. Although the bulk of total output is concentrated in a relatively few large firms no single producer accounts for as much as 5% of the annual tonnage. Entry into the industry is easy, so that there are always literally thousands of small mines operating on the fringe, constituting a persistent threat to the stability of the price structure.

What is more important, the demand for bituminous coal is highly sensitive to changes in the level of gross national product. Through the principle of acceleration, relatively small adjustments in aggregate demand in the economy at large will cause more than proportionate changes in the demand for coal. The smallest of cyclical or seasonal fluctuations in general economic activity will be transmitted to the bituminous industry, and the effects will be magnified. A powerful element of instability is thereby built into the coal industry.

Of at least equivalent importance, there has been over the last thirty years a pronounced secular change in coal demand. Although total consumption of all fuels has expanded tremendously since 1920, consumption of bituminous was at about the same level in 1950 as in 1920. In other words, the demand for bituminous coal has diminished relative to substitute sources of energy.

While conditions of demand have been the critical element, the industry's position has in the past been complicated by intense interregional competition, particularly involving mines in the Appalachian area north and south of the Ohio River. The chief weapons in the fierce struggle for markets between northern and southern firms were freight rates on the one hand and wage rates on the other. It was through the latter, of course, that the United Mine Workers' interests were directly involved, since in coal mining there is an intimate relationship between wage rates and labor costs per unit of output and between labor costs per ton and total producing costs per ton.

As I shall show later, the period of intense interregional competition was a traumatic experience for the United Mine Workers. It is safe to say that the present policy of the UMW is based predominantly on the union's determination that such a situation shall not recur. This has meant that it has arrogated to itself the largest share of the responsibility for stabilizing the industry.

The central problems considered in this study, then, are three: First, what does the union want? Second, what methods does it employ to achieve its objectives? Third, what have been the effects of union policy on the bituminous industry and the national economy?

Broader Issues

Many of those who read this book will, I suspect, have formed in advance their answers to the questions just posed. This is understandable. So great has been the notoriety of the miners' union in recent years, and so prolific have been the popular commentators on the subject, that much of the essential data about the coal industry is well known. It may seem to some, therefore, that I have done no more than assemble the relevant information in one volume, taking care to surround the statistics with suitably erudite observations.

Let all be assured that this work is conceived on a broader scale than one of mere reporting. There has been no attempt to labor the obvious, e.g., have the increasing wage rates of the miners contributed importantly to changes in production costs per ton and in product prices? (They have.) Rather, a serious effort has been made to deal with more general questions, questions of significance to the study of the entire labor movement. What effect has unionism in the coal industry had on the allocation of resources? Have union wage policies affected the structure and behavior of the market for bituminous coal? To what extent is the union responsible for the substantial wage gains that have been made by employed miners since 1933?

Issues of these kinds are not easily resolved, since the body of data is in each case susceptible to varying interpretations. Thus, for example, there are some economists who argue persuasively that unionism represents a serious threat to the successful functioning of the competitive enterprise system; that, on the one hand, it interferes seriously with the functioning of the pricing mechanism and, on the other, it is an engine of inflation. Yet there are equally reputable economists who maintain that unions have little effect on wage rates over a period of years, that the available evidence suggests instead that trade unions are little more than a transmission belt through which market-determined wage rates are passed on to union members.

It is extremely doubtful that the conclusions I have reached on these issues will meet with universal acceptance. This is clearly no excuse, however, for avoiding an honest effort to deal with them, and such an effort has been undertaken.

It is scarcely necessary to say that union policies and their effects have been conditioned by the economics and politics of the bituminous coal industry. Five chapters, consequently, have been devoted to these matters. In Chapter 1 is a brief résumé of the market structure and behavior of the industry, including the number and size of firms, the nature and behavior of costs, the conditions of demand, the pricing process, and similar matters. There follows in Chapter 2 a description of the Appalachian coal region and the relationships among the competing coal fields contained in it. Significant trends in the industry from 1890 to 1950 are reviewed in Chapter 3, with a view to placing the present attitudes of the miners and operators in historical perspective. Then in Chapter 4 union policies during the same six decades are examined, with special emphasis on the UMW's policy objectives since 1933. Finally, in Chapter 5 the political aspects of union-management relationships are explored.

The core of the book is contained in the concluding four chapters. The effects of union policy on the level of wage rates is discussed in Chapter 6. The union's impact on competitive relationships among coal fields in the Appalachian area is examined in Chapter 7, and its effect on interfuel competition in Chapter 8. The concluding chapter is devoted to an appraisal of the motivations of the union, the long-run effects of its strategies, and the problem posed thereby for public policy.

CHAPTER 1

Market Structure and Behavior

THE NUMBER AND SIZE OF FIRMS

THE COMPETITIVE STRUCTURE of the bituminous coal industry is indicated by the fact that in 1948 roughly 8,000 mines were operating, each producing 1,000 tons or more annually.[1] There was, in addition, a large but uncounted number of mines with an annual output of less than 1,000 tons. Not all active operations were competing with one another, however, since some enterprises operated a group of mines. In 1943, for example, 128 financial groups controlling 272 coal companies produced more than 60% of the national output.[2]

For some years there has been a tendency in the industry toward larger firms. In 1905 approximately 40% of the national tonnage was produced by firms mining 200,000 or more tons per year. By 1947 the proportion had increased to 66%. Even so, the very small mines—those producing 1,000 to 10,000 tons per year—increased their share of total output from 2% in 1905 to 3% in 1947.[3] To some extent the latter figures represent the hardiness of firms interested chiefly in serving local or neighborhood markets, but the data also indicate the presence during 1947 of marginal operators seeking their share of the prevailing high profits.

The industry, in sum, has many firms, and despite increasing concentration of output, no one firm or group of firms is dominant.[4] It takes 128 control groups to account for 60% of total tonnage, and there is a host of small mines which poses a constant threat to price stability in the industry. A further limitation on the power of the large firms arises from the relative ease of entry into bituminus coal mining.

1. U. S. Bureau of Mines, as reported in National Coal Association, *Bituminous Coal Annual, 1949* (Washington, 1949), p. 47.
2. See Table 1-1 in appendix.
3. National Coal Association, *Bituminous Coal Annual, 1949,* pp. 79, 42.
4. The largest single firm, Pittsburgh Consolidation Coal Co., operated 39 underground and 9 surface mines in 1946. Its output in all mines amounted to less than 5% of the national tonnage. Ibid., p. 78.

Conditions of Entry

A major reason for the large number of firms in soft-coal mining is that it is not difficult to enter the industry. Ownership of land in which there are coal deposits is widely scattered, making the operation of a mine possible for a large number of independent persons. The principal obstacle to entry is the initial investment outlay, which varies according to the scale of mining operations planned.

Coal mining is of three general types: deep or underground, surface or strip, and punch mining. Investment requirements for each type of operation decline in the order named.

A deep mine requires a "tipple" or platform on which the loaded mine cars are tipped over, their contents dropping on sizing screens. In modern deep mines, equipment is also provided for washing the coal. A modern tipple, complete with electrically controlled washing, full screening, and capacity to last 20 to 25 years, represents an outlay of between $500,000 and $2,500,000.[5]

Mining machinery, Airdox or Cardox systems for taking the coal down, transmission lines, fans and blowers, machine shops, stores, and the many other items that go to make up a good coal mine operation cost another $500,000. To open a deep mine which is planned for long-term operation, then, total outlays of $1,000,000 to $2,000,000 are required. Put another way, a modern mine involves the expenditure of not less than $5 to $6 per ton of annual capacity—and that under the most favorable conditions. In less favorable locations the per ton investment cost will rise as high as $12.

Strip mining, meaning surface operations, is substantially easier to enter on short notice. Firms which plan to remain in the industry on a long-term basis may be required to invest initially up to $1,000,000, or about $5 per ton of annual capacity. The outlay will include the purchase of heavy trucks and trailers, large power shovels or drag lines for the removal of overlying strata, and smaller rotary shovels for the removal of the coal.

Smaller stripping operations, however, necessitate relatively little initial investment. Outcroppings of coal can be worked with power shovels of the size employed in highway construction. In fact many small strippers were originally road builders who were delighted to employ otherwise idle machinery for coal mining. Other strip operators can rent such machinery at relatively little expense. Strip mining can be undertaken on very short notice for as little as $10,000. That low

5. A substantial part of this spread in total outlay is represented by different prices of coal lands in various parts of the country.

figure explains in large part the rapid growth of strip mining after 1920 and its continued expansion to the present.

Least expensive of all is the punch operation. This type of mining was developed to reclaim coal that would not be recovered in the process of regular deep-mine operation. Most of the punch-mining operations take place in deep mines which are nearly or completely exhausted. Workers simply blast out that coal which was left untouched by earlier operators. Loading is by hand, as is haulage of the product to the surface. Undertaken principally by families or friendly groups, both of which usually operate as nonunion workers, punch mining requires little more than a small truck, a little dynamite, and some initiative. Total initial outlay may be as low as $1,000.

Summing up, the intended output and degree of permanence of a coal mine determines its initial investment cost, beginning at perhaps $1,000 for a punch mine, and from $10,000 for a strip mine up to $1,000,000 for a permanent large-scale strip-mining tract. For deep mines producing high-quality washed coal, initial capital requirements run between $1,000,000 and $2,500,000.

Where relatively low initial investment outlays are required, entry into the industry is virtually unrestricted. This means that prospective entrants are highly responsive to changes in prices and expectations of profits. Before and after both World Wars there were marked increases in product prices which were accompanied by substantial accruals to the output potential of the industry. Even when existing physical capacity is more than adequate to meet prevailing conditions of demand, relatively small increases in prices will attract new entrants. Such was the case in 1926, for instance, when a strike abroad caused a mild boom in American exports of bituminous.

It has been argued that heavy taxes and carrying charges "compel owners to seek immediately such income as can be derived by exploitation even though the market would not otherwise justify immediate development." [6] While this argument may explain why *existing* mines continue to operate even when total receipts are less than total production costs, it does not describe the forces which impel a coal operator to open a new mine. As one coal operator has put it, "The factor that is most forceful in getting coal lands developed is a strong coal market, which practically promises a high rate per net ton of coal sold." [7] It seems highly doubtful that fixed charges could compel firms

6. R. H. Baker, *The National Bituminous Coal Commission* (Baltimore, Johns Hopkins Press, 1941), p. 21.

7. Personal letter to author from G. H. Seal, Vice-President, C. H. Sprague & Son Coal Co., Boston, May 11, 1951.

to enter the industry. Entry is undertaken only if the operator expects high returns—enough to cover operating costs plus a reward for spending the funds to open the mine.

"The cost of carrying undeveloped land is not of sufficient expense to force the coal land-owner to enter the industry in any great rush. How could there be heavy fixed expenses if the land had not been developed? The taxes on undeveloped coal lands are not heavy, and it is not until development starts that any overhead or fixed expenses begin to amount to any large figure." [8]

Conditions of Exit

As in other industries, coal men are reluctant to close up shop, even temporarily. They maintain a perpetual air of confidence about the future. They can be likened to the motorist who insists that accidents always happen to the other fellow. More than mere self-confidence accounts for this attitude. Few businessmen are willing at the first provocation to give up their positions of power and prestige. To the limit of their financial ability they will try to ride out the storm. Moreover, the experienced hands in the coal industry are well aware that product demand fluctuates over a wide range in comparatively short periods of time. A war, a strike, a railroad car shortage—any of these can turn a depression in the coal trade into a comfortable prosperity. For all these reasons coal mines may be kept in operation even when financial losses are substantial.

Continuation of production even when product price is less than out-of-pocket costs is sound business policy under certain conditions. In this context, as suggested above, it may well be correct to assert that heavy taxes and carrying charges "compel owners to seek immediately such income as can be derived by exploitation even though the market would not otherwise justify [it]." From a practical standpoint

> it can definitely be stated that an operator will not close a mine down as soon as he fails to cover his "out-of-pocket" expenses. . . . The operator has 100 to 400 employees, who are dependent upon his mine for any kind of livelihood, and should the mine close up, the operator stands to lose a considerable amount in accounts receivable at the [company] store, and also stands to lose his entire working force who will have to depart for greener pastures.[9]

In theoretical terms the effect of carrying charges on production decisions involves the concept of user cost. In the words of J. M. Keynes, user cost is

8. Ibid.
9. Ibid.

the reduction in the value of . . . equipment due to using it as compared with not using it, after allowing for the cost of the maintenance and improvements which it would be worthwhile to undertake and for purchases from other entrepreneurs. . . . It is the expected sacrifice of future benefit involved in present use which determines the amount of the user cost, and it is the marginal amount of this sacrifice which, together with the marginal factor [input] cost and the expectation of the marginal proceeds, determines [the entrepreneur's] scale of production.[10]

User cost varies directly with the entrepreneur's expectations concerning the future level of profits; if he expects that product prices or costs will remain at their present levels for a protracted period of time, the "expected sacrifice of future benefit," or user cost, will be low. And, as Keynes pointed out, "very low user cost at the margin is . . . a characteristic . . . of particular situations and types of equipment where the cost of maintaining idle plant happens to be heavy. . . ."[11]

The concept of user cost permits an explanation for the apparent slowness with which coal mines are withdrawn from production when product prices are falling. A modern mine is equipped with a substantial amount of mining machinery, including devices for drilling, cutting, and loading. Large money outlays will have been made for intramine haulage equipment, such as mine cars, locomotives, conveyor belts, and rubber-tired tractors and trailers or their equivalent. A vast network of electrical transmission lines will have been constructed. An elaborate ventilation system will have been installed. Unless the intent of the operator is to abandon the property completely, precautions must be taken against deterioration of the mine: water seepage must be controlled; accumulations of marsh gas must be dispelled; roofs must be maintained to prevent cave-ins. And, of course, property taxes and insurance charges must be paid.

If the marginal user costs of this plant and equipment are low, the entrepreneur has a strong motive to maintain a high level of output. For with low marginal user costs the present value of output exceeds the (discounted) value of future output. When it is recognized that there is also a user cost on the unrecovered coal in the mine and that this cost will also be low when the entrepreneur is pessimistic about future prospects of prices and costs, it is clear why his scale of production in the present will be kept at high levels.

User cost does not alone account for the tendency of output to remain high when product prices are relatively low. Much depends on the cash asset holdings of the entrepreneur. If his liquidity position is

10. J. M. Keynes, *The General Theory of Employment, Interest and Money* (New York, Harcourt, Brace, 1936), p. 70.
11. Ibid., pp. 72–3.

strong his user cost will be raised, since the rate of discount used by the coal operator will be relatively low. Under these conditions the owner will be constrained to conserve his assets. Firms with poor liquidity positions, such as marginal or "fly-by-night" companies, will on the other hand be impelled to maintain high levels of output. Their user cost will be low since lenders will impose upon them a high rate of discount. And they must strive to maximize their cash receipts in order to meet such inescapable expenses as royalties, property taxes, and so on. Unless product prices continue to decline, mining operations can continue for some time. If the point is reached where it is no longer possible to meet the unavoidable expenses out of current receipts, the operator must then withdraw.

When things get *so* tough, the operator tries to find someone with greater optimism who will purchase the property. In this manner an operation drags along during a dull period until eventually the dividend for coal increases. The old operator who has stayed with his losing proposition may begin to recoup his previous losses—or some of them. A new operator who has taken over finds he has made a pretty good deal, for a short time at least.[12]

The Problem of Excess Capacity

It is commonly concluded that the bituminous coal industry's problems derive from the persistence of excess capacity. This is a typical comment: "The problem of bituminous coal . . . may be considered as focusing on three factors: (1) excess capacity against a limited market, resulting in (2) intense price competition to hold a share in the market, resulting in (3) the breaking down [sic] in the wage structure in order to obtain lower costs and preserve existence." [13]

There is an unmistakable implication in the quotation that excess

12. G. H. Seal, loc. cit.

13. F. E. Berquist and Associates, *Economic Survey of the Bituminous Coal Industry under Free Competition and Code Regulation* (Washington, U. S. National Recovery Administration, 1936), *1*, 61. It should be noted that this work, among others, uses the Bureau of Mines index of capacity as a measure of excess capacity. The following is pertinent: "Overcapacity cannot be identified with unutilized capacity in any simple or unique sense, although it is common in discussions of the problem to do so. The Bureau of Mines index of capacity . . . indicates how much coal would be produced from existing mines if those mines were worked one normal shift on every working day of the year, and if the further exploitation of the mines resulted in uniformly proportional increases in output. The index thus reports that in a particular period the existing number of mines could have produced more than was produced. If other conditions are chosen as significant—if, for example, one measures how much could be produced by working the mines two shifts a day, or on Sundays—the index would be quite different, but would have exactly as much or as little significance for prices as it has now." E. V. Rostow, "Bituminous Coal and the Public Interest," *Yale Law Journal, 50* (1941), 550.

capacity causes price cutting. Such an assertion is of limited accuracy at best. Overcapacity is more a result than a cause of falling prices, as a careful definition demonstrates. Excess capacity exists

> if the quantity of [productive] factors associated with the industry would have been less if the present prospects of future demand and costs had been foreseen; or if in view of the present prospects, fixed factors show a desire and intent to withdraw from the industry with the passage of time . . . The existing supply curve of the industry over the relevant price range is to the right of the position in which it would be if sufficient time were allowed for adjustment to present expectations.[14]

As will be discussed in detail later, the coal industry was burdened with excess capacity during the 1920's and 1930's. This was a result largely of the end of the secular rise in consumption of bituminous coal after World War I. There was, too, a prolonged period of intensive price competition based on differentials in wage and freight rates. For many mines operations became exceedingly unprofitable. Between 1923 and 1932 in particular there was a gradual but steady withdrawal of resources from the industry.

There is some validity to the argument that wage rates, for example, would not have been so severely depressed during the twenties had surplus labor withdrawn from the industry more rapidly than it did. Mine labor, like mining machinery, was relatively immobile because it had few alternative employment opportunities. As a consequence, labor was willing to remain in the industry even though wage rates declined.

An excess of productive factors relative to present and future prospects of demand and costs develops, in sum, because of the immobility of productive services in the face of fluctuating product demand. New firms are attracted into the industry by the existence of "profits" (economic rent). Since entry is relatively easy, total supply is increased quickly when the new entrants begin production. Unless product demand continues to expand, however, the price structure soon weakens. Profits begin to decline.

In time the awareness of falling profits or increasing losses becomes general. Some mines, especially those with high costs relative to their competitors, will withdraw. But many firms, including those with low user costs and a poor liquidity position, will continue to operate.

Since the net returns from operating are greater than the net returns from not operating, the survivors will try to operate on a full

14. J. P. Miller, "Pricing of Bituminous Coal. Some International Comparisons," in *Public Policy*, ed. C. J. Friedrich and E. S. Mason (Cambridge, Harvard University Press, 1940), *1*, 151.

work day and work week.[15] Total supply stays high, while demand remains unchanged or declines. As prices tumble there begins a frantic search among the operators for a means of reducing costs, particularly labor cost, which represents upward of 60% of total outlays. Unless there are floors under wage rates, whether maintained by the union or the government, a downward price-wage spiral will result.

Gradually the high-cost firms will be eliminated from the industry. If there are no alternative opportunities for the displaced resources, they form "intractable pools of unemployment." [16] Otherwise they drift into other industries, perhaps permanently to remain away from coal mining.[17]

The Nature and Behavior of Costs

Costs of production differ widely among firms and regions in the bituminous industry. The factors accounting for the differences include: [18]

1. Geological conditions
2. Degree of mechanization
3. Wage-rate differentials
4. Type of mining (surface or underground)
5. Age of mine and method of mining (advance, retreat or "pillar robbing")
6. Differences in length of work day and work week
7. Differences in amount of coal preparation
8. Efficiency of labor and management
9. Age of mining equipment
10. Depth of mine
11. Rate of employee compensation insurance
12. State and local taxes
13. Amount of mine acreage being worked

Most of the items on the list are self-explanatory. A few could be discussed at some length only if the intent here were a full-dress study of the economics of coal mining.[19] Since that is not the case, only those factors judged to be most important will be considered at any length. They include: differences in costs arising out of geological

15. The nature and behavior of costs are discussed below.

16. The phrase is from C. Glasser, "Union Wage Policy in Bituminous Coal," *Industrial and Labor Relations Review* (July 1948), pp. 609–10. It partially describes the crisis in coal that extended from 1923 to 1932.

17. This may well have happened after 1940. See Chap. 7.

18. Adapted from National Bureau of Economic Research, *Report of the Committee on Prices in the Bituminous Coal Industry* (New York, 1939), pp. 21–5.

19. This applies, for example to the age of the mine and the method of mining.

conditions, alternative types of mining, differences in operating time, and the degree of mechanization. The effects of wage-rate differentials will be considered in detail in a later chapter.[20]

The nature of cost differences among regions is depicted in Chart 1.[21] The variations in total producing costs per ton range from a high of $3.39 in District 7 to a low of $2.50 in District 4. The difference between the two extremes was about 90¢ per ton.[22]

Unit production costs are relatively high in Districts 1, 2, 7, and 8. Mines in those districts have, in general, coal seams which are thinner than those found elsewhere in the Appalachian area. This has a significant effect on costs of production.

The coal seam of moderate thickness presents the least mining difficulties. Any decided thinning limits production, decreases recovery and therefore increases costs. Thickening of the seam in underground mining has the same effect; for, as the seam increases in thickness, it becomes necessary to maintain larger pillars, timbering becomes more difficult and finally impracticable and roof control is almost impossible. The limited information available indicates that for maximum recovery in underground bituminous coal mines the ideal seam thickness lies between 6 and 8 feet.[23]

The available data [24] indicate a heavy concentration of output in seams two to four feet thick in Districts 1, 7, and 8. Those were the districts, of course, with highest unit production costs. Only in District 2 among the high-cost districts was there no concentration of production in the thin seams. In Districts 3, 4, and 6, on the other hand, most of the coal was mined from seams with a thickness of four to eight feet. Their superior cost position relative to that of their competitors is readily understandable.

The impact of poor geological conditions is far-reaching. In low seams, where the coal vein is less than three feet high, standard-size mining machinery cannot be used. If the mine is to be mechanized, specially constructed locomotives and mine cars must be purchased. Unless the mine operator has large cash assets or a good source of credit, the heavy money outlays which may be required will preclude

20. See Chap. 8.

21. The same data in tabular form may be found in Table 1-2 in appendix. Regions included in the several districts are identified in Chap. 2 where the entire Appalachian area is defined and described.

22. The data given are averages for all kinds of mines—hand loading, machine loading, and stripping. The figures are, as a result, averages of average producing costs for all mines of each type. Because some important distinctions as among the kinds of mines are thereby concealed, see Tables 1-3, 1-4, 1-5 in appendix.

23. W. H. Young and R. L. Anderson, *Thickness of Bituminous Coal and Lignite Seams Mined in U.S. in 1945,* Information Circular 7442 (U. S. Bureau of Mines, 1947), p. 2.

24. See Table 1-6 in appendix.

CHART 1

Average Producing Costs for Commercial Mines, All Types, 1945

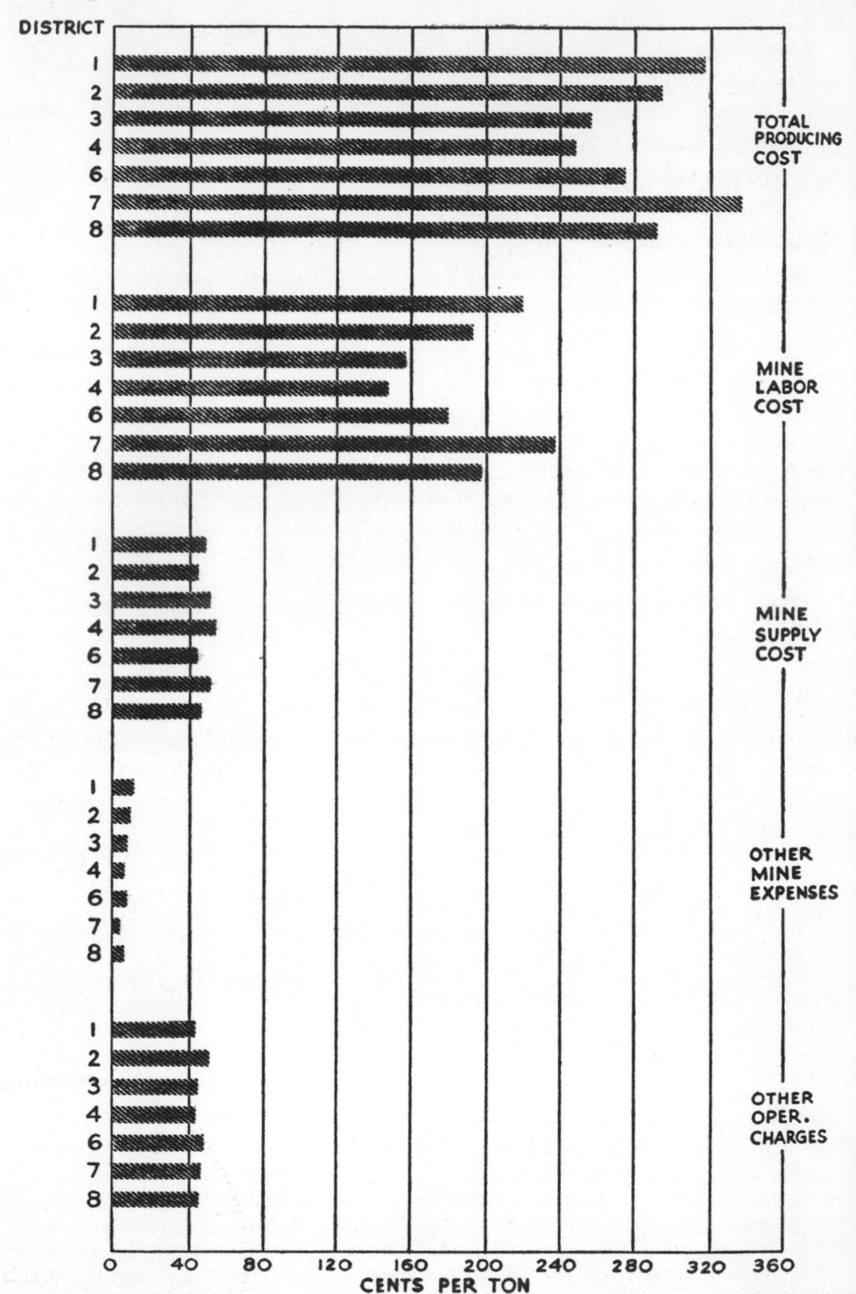

Source: Office of Temporary Controls, *OPA Economic Data Series,* No. 2 (Washington, [1946?]), p. 7

mechanization. Inability to mechanize may well magnify an already unfavorable cost position relative to one's competitors.

Unfavorable mining conditions tend to reduce productivity of labor and thus raise unit production cost.[25] If the seam is low, the miner may be required to work from a prone position, a particular handicap in hand-loading mines. If the coal is interwoven with slate and similar impurities, the miner's output is sharply reduced by the effort he must spend in "gobbing" (separating the impurities prior to loading the broken coal).

Productivity is, of course, a function of the efficiency of the mechanical equipment in use. Several types of loading machines are available, each designed for use under varying conditions.[26] They differ in efficiency and in the number of workers required for operating them.[27] Favorable resource conditions permit operators of deep mines to use the most efficient equipment. The machines maximize the benefits of the good resource conditions. The two, in other words, mutually support each other, substantially reducing costs of production. These factors account in the main for the differences in production costs among competing coal regions.

There are three different ways to open an underground mine. The method adopted depends principally on the terrain surrounding the coal vein. *Drift mines* are opened by driving horizontally into the coal seam; such mines are generally located on the side of a hill. *Slope mines* are those which are opened by tunneling into a hillside at an incline. *Shaft mines* are those which reach the vein by a vertical descent through the overlying earth strata. Such mines are typically located in flatlands or plains.

25. See Table 1-7 in appendix for statistical details.

26. The *pit-car loader* is a device with a hod-like attachment which can be lowered to the floor of the mine. Coal is loaded into the elevator which conveys the coal into the mine car. The *scraper* shovels the broken coal into a conveyor belt connected to the mine car. It is useful only in mines where timbers are not too close together, since freedom of action is essential. Moreover, the mine floor must be hard lest the scraper dig up impurities while it is operating. The *duckbill,* or self-loading conveyor, is a shovel-like device with a flared mouth, which is attached to a shaking conveyor. The differential movement of the conveyor carries the coal from the loading mouth to the mine car. It has the same limitations of use as the scraper. *Hand-loaded conveyors* are attached to an endless belt conveyor ("Mother") which transports the coal either to the main haulageway to be dumped into mine cars or to another conveyor which carries the coal directly to the tipple. They are of greatest value in places where the other devices are impracticable. The *mobile loader* is of two types, having either a shovel attachment or gathering arms. Conveyors load the coal directly into mine cars. It is by far the most efficient machine in terms of tons loaded per hour. But it is of little use in thin seams or where there is close timbering.

27. For a measure of the comparative effectiveness of the machines, see Tables 1-8 and 1-9 in appendix.

Once the coal seam is reached, long entries or corridors are developed. In almost all American deep mines "rooms" or small working areas are developed off the entries. Pillars are left standing to support the roof. Thus, the American system is known as the room-and-pillar system.

Various precautions in addition to roof support must be taken. Adequate ventilation must be provided to assure a continuous inflow of fresh air and outflow of dangerous gases, chiefly methane. Coal dust must be subdued to avert explosions. Electrical transmission lines must be erected for lighting and communications. Transportation must be provided to insure rapid and efficient transit of men and coal.

Actual mining procedure is relatively simple, at least in description. The coal "face" is first undercut or sheared vertically [28] to render the blasting more safe and more productive of lump sizes, since "shooting off the solid," or blasting with no prior cutting or shearing, results in greater breakage of the coal. It also puts greater strain on roofs and pillars, increasing the danger of collapse.

After the cutting, holes are drilled into the face for the insertion of blasting charges. The blast, of course, shakes down the face of the seam. Not all the loosened matter will be coal, for impurities such as slate are commonly encountered in the vein. Some of the impurities may be gobbed, or pushed aside, prior to loading; the remainder will be removed at the coal washery. Whatever the case, the coal is loaded into haulage equipment and transported to the preparation facilities at the mine mouth.

Strip mining, while involving the same basic procedure of blasting and removal, is a somewhat different kind of operation. It is common in areas where the coal vein is close to the surface. The "overburden," or overlying earth strata, is removed with the aid of earth-moving equipment, including bulldozers, draglines, and power shovels.[29] Once the vein is laid bare, blasting is employed to loosen the coal. Rotary power shovels then load the broken coal into haulage equipment (usually trucks) which removes the coal to preparation facilities.

Strip mining has certain advantages over various types of underground operations. Timbering is unnecessary, as is ventilation equipment. Haulage units are not limited in size by the height or width of the mine. Strip mines can be opened and put into production far more rapidly than underground mines, since preliminary preparations are much less extensive and time consuming. This latter consideration

28. This process is known as "kerfing."

29. Small family-run mines, working outcroppings of coal along highways or railroads, may be worked totally by hand, but such mines account for only a tiny part of total output.

results, of course, in more rapid recovery of fixed charges. Furthermore, strip-mining equipment tends to have more alternative uses than does deep-mining machinery, thereby reducing the cost of exit from the industry. And strip mining, a relatively recent development, "has been less restrained by the hand of tradition, which rests heavily on underground mining and [the strip operator] has been free to mechanize as the opportunity arose." [30] Finally, and most important, far less labor is needed in surface than in underground mining.

CHART 2

Average Producing Costs for Commercial Mines, By Kind of Mine, 1945

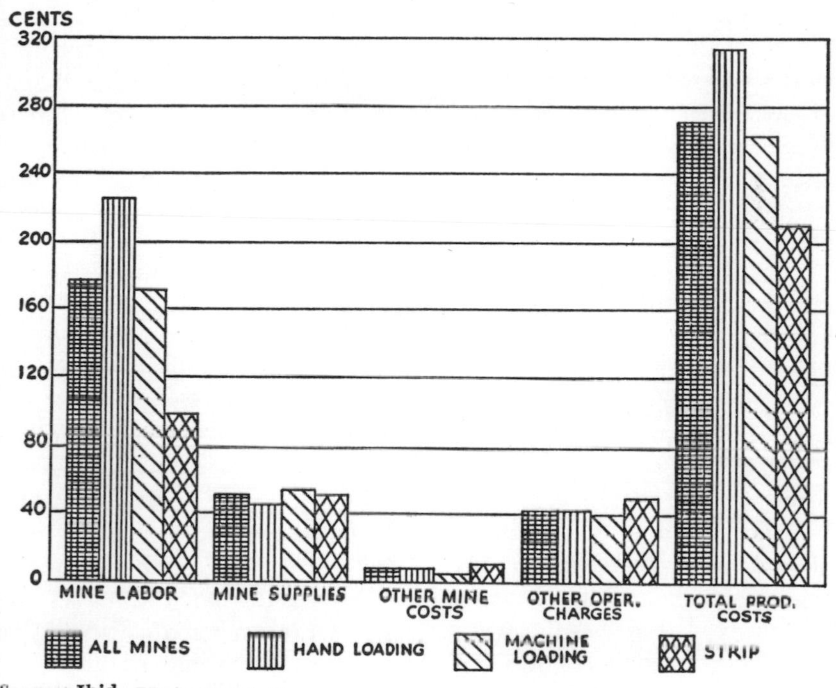

Source: Ibid., pp. 7, 13, 19, 25

In summary, unit cost in strip mining is substantially less than in underground mining.[31] The advantage of the surface mines is centered almost entirely in unit labor cost. The favorable differential in that component of cost is, in fact, partially offset by higher mine supply and other operating charges in the surface mines. As is to be ex-

30. U. S. Bureau of Mines, *The Economics of Strip Coal Mining,* Economic Paper 11 (Washington, 1931), p. 6.

31. For a comparison of producing costs in underground and surface mines, see Chart 2.

pected, administrative and selling expenses are virtually the same for both kinds of mines.

As between hand- and machine-loaded operations, a similar situation obtains. Average cost in machine-loaded mines averages 55¢ per ton less than in hand-loaded mines; the unit labor cost differential is 60¢ per ton. Unit cost of mine supplies, however, is lower in hand-loaded mines, partly counterbalancing the sizable differential in unit labor costs.

Mining costs may be classified as avoidable and unavoidable.[32] The first are costs which can be adjusted as the level of output of the firm varies. They include in the short run: labor costs other than supervisory; maintenance supplies for mining machinery; purchased power and coal used to produce coal; other mine expenses, such as hired trucking to the tipple, ramp, and so on; unemployment and workmen's compensation taxes. Unavoidable costs include all expenses which are incurred whether the mine produces one or 100,000 tons of coal. They include in the short run: prepaid labor costs, consisting principally of payments to employees on a salary or fairly long-term (weekly or monthly) wage basis; property taxes; insurance premiums; amortization charges; depreciation expense.

The average avoidable cost (AAC) function is dish-shaped. At low levels of output, total avoidable costs increase at a slower rate as production expands than does output itself. Average avoidable costs decline, a result in large part of the fact that indivisible productive services such as labor are not being used at maximum efficiency. Output may rise from very low levels, that is, by the more efficient utilization of the resources already associated with the firm.

As the volume of output expands, more resources must be employed. Since the supply curve of factors of production facing the representative coal firm tends to be highly elastic, more resources can be employed at no significant increase in average avoidable costs. The AAC function, as a result, is constant over a wide range of outputs.

Relatively constant AAC extends to a definite limit, after which it will rise sharply. The limiting point is, of course, that of full utilization or capacity.[33] Increments to output after capacity will necessitate either the hiring of additional resources or the more intensive

32. For a lucid analysis of this distinction see W. A. Lewis, *Overhead Costs* (London, Allen and Unwin, 1949), chap. 1. The discussion here is confined to the short run.

33. Just where the point of capacity operation is located is uncertain. It varies among firms, some considering the optimum point as a four-day week, others a five-day week.

use of those productive services currently employed. In either case money outlays will tend to rise more rapidly than does output.

Average unavoidable costs behave quite differently. Since unavoidable costs are by definition unrelated to changes in output, increased production will cause a decline in unavoidable costs per ton. The unavoidable cost function is reversible; as working days per month are reduced, unit unavoidable costs rise rather sharply.

> Taking a month of 25 working days as full-time operation, it is found that when the mine works 16 days, four days a week, the cost per ton is increased 8 to 9 per cent over full-time operation; when working time is 12 days per month or 3 days per week . . . the unit cost is 21 to 25 per cent over full time or minimum cost; and where but 8 days per month, or 2 days per week, are worked, costs increase 48 per cent. . . .[34]

It is clear, therefore, that the average unavoidable cost curve descends rather sharply as output increases from very low levels, but tapers off as output continues to increase up to capacity. By adding the unit avoidable and unavoidable cost curves, an average total cost curve may be derived. This curve will have a dish shape. It will be falling at low levels of output, be constant over a wide range of outputs, and then rise again as capacity is passed. The flatness of the middle of the curve is, of course, a result of the relative constancy of the average unavoidable cost curve.

Evidence attesting to the accuracy of the curves described above is presented in Tables 1 and 2. Though it may appear on the surface that the two tables show the same things, both are necessary to make the point. Mining conditions differ among regions, giving rise to important cost differences.[35] Table 1 shows producing costs in eastern Pennsylvania, where coal seams are rather thin and full of impurities. Table 2 includes costs in northern West Virginia, where coal veins are thick.

Both tables indicate a decline in average total cost as the number of days or shifts worked is increased. Moreover, the reductions in unit cost are greatest with an eight-hour day. The decline in all cost components is continuous, however, reaching its low point in the six-day, eight-hour work week. This seems to conflict with the earlier observation that avoidable and total unit costs rise after capacity. The difficulty, of course, is that the wage rates were assumed to be constant for any number of days worked. Had overtime rates been admitted into the calculations, it is likely that per ton labor cost in a

34. U. S. Coal Commission, *Report on the Effect of Irregular Operation on the Unit Cost of Production* (Washington, 1925), Pt. 8, p. 1975.
35. See above, p. 9.

six-day operation would show an increase rather than a continuing decline. Since labor costs are so large a proportion of total costs, an upward turn in the former would certainly have caused a like movement in the latter.

Table 1. Effect on Producing Costs of Coal Mining of Variations in Number of Hours and Days Worked in a Typical Deep Mine
(Base period: 5 days, 8 hours) *

Hours Worked	*6 Days*	*5 Days*	*4 Days*	*3 Days*
	TONS OF OUTPUT			
8	3750	3125	2500	1875
7	3281	2734	2187	1640
6	2812	2344	1875	1405
	INCREASE IN PER TON LABOR COST			
8	$—.029	$ 0	$.044	$.118
7	—.029	.001	.046	.121
6	—.028	.003	.014	.037
	INCREASE IN PER TON SUPPLY COST			
8	—.005	0	.006	.022
7	—.005	.001	.011	.027
6	—.004	.003	.014	.037
	INCREASE IN PER TON OVERHEAD COST			
8	—.080	0	.121	.322
7	—.023	.069	.207	.438
6	.054	.161	.322	.607
	TOTALS			
8	—.114	0	.171	.462
7	—.057	.071	.264	.586
6	.022	.167	.385	.769

* Costs are based on existing average hourly wages in the area where the mine is located. The data used here are for a mine in Tioga County, eastern Pennsylvania, in November 1933.

Note: Overtime payments are not included in the cost figures, since the prevailing contract in 1933 did not so provide.

Source: U. S. National Recovery Administration, *Report of Mining Engineers Appointed . . . to Investigate the Practicability . . . of Applying to Bituminous Coal Mining a Shorter Work Day and Work Week* (Washington, 1934), p. 5.

Table 2. Effect on Producing Costs of Coal Mining of Variations in Number of Hours and Days Worked in a Typical Deep Mine (Base Period: 5 days, 8 hours) *

Hours Worked	6 Days	5 Days	4 Days	3 Days
	TONS OF OUTPUT			
8	8400	7000	5600	4200
7	7350	6125	4900	3675
6	6300	5250	4200	3150
	INCREASE IN PER TON LABOR COST			
8	$—.012	$ 0	.020	.051
7	—.012	.002	.021	.055
6	—.011	.003	.025	.060
	INCREASE IN PER TON SUPPLY COST			
8	—.008	0	.011	.029
7	—.006	.002	.016	.036
6	—.005	.004	.018	.042
	INCREASE IN PER TON OVERHEAD COST			
8	—.048	0	.067	.182
7	—.012	.037	.116	.248
6	.031	.091	.182	.334
	TOTALS			
8	—.068	0	.098	.262
7	—.030	.041	.153	.339
6	.015	.098	.225	.436

* Costs are based on existing average hourly wages in the area where the mine was located. The data used here are for a mine in northern West Virginia in November 1933.

Note: Overtime payments are not included in the cost figures since the prevailing contract in 1933 did not so provide.

Source: Ibid.

This limitation aside, it seems fair to conclude that the curves postulated above are essentially accurate. Average total costs decline rapidly at first, then level off over a broad range of output. Marginal cost is, in other words, constant from fairly low volumes of output to the point of capacity. This means that the mine operator has a strong incentive to maintain a high level of output as long as marginal revenue exceeds marginal out-of-pocket costs. He may be minimizing his losses, that is, by producing at capacity rather than by operating on only a part-time basis. This helps to explain why output has remained high in the industry even during periods of falling prices. There is that suggestion in the following remarks of a prominent coal operator:

Question: Mr. O'Neill, why did you continue operating your mines under conditions [of competitive wage and price-cutting]?

Answer: Well, you get to a point when you must do something, you have an investment of $3 or $4 a ton (of) annual output . . . with modern equipment. If you have a 100,000 ton mine you have an investment of $300,000 to $400,000, and that is a small operation. You cannot let a mine stand idle. . . . In our district it will cost 20 to 30 cents a ton of the annual capacity to leave the mine stand idle wholly and take care of the property and pay its expenses. In other words, if you have a 50,000 ton a month mine, it will cost you $10,000 to $15,000 a month to let it stand idle. And that is the pressure that is in back of an operator, to try to operate his mine, to try to accumulate enough tonnage in order that he can reduce those losses, and he cannot do it by idle time.[36]

The Nature of Demand

The demand for coal is a derived demand. A list of the principal users of bituminous reflects the dependence of coal operators on buyers other than ultimate consumers. What is more important, most coal sales are made to producers of capital goods. In 1950, for example, only 19% of the tonnage consumed was delivered to retail dealers. The remainder was divided as follows: coke, gas, and steel producers, 25%; other industrials, 23%; electric utilities, 19%; and railroads, 14%.[37]

Derived as largely as it is from the demand of producers of capital goods, demand for bituminous is highly sensitive to changes in the level of economic activity and is thus characterized by marked cyclical instability. By the principle of acceleration small variations in consumer demand cause relatively larger changes in the derived demand for productive services. The sensitivity of coal demand to adjustments elsewhere in the economy was clearly demonstrated during the recession of 1948–49, when gross national product declined by less than 5% while consumption of bituminous fell by 27%.[38]

36. Testimony of C. O'Neill, *Carter v. Carter Coal Co.,* Verbatim Transcript of Proceedings before Supreme Court of District of Columbia (National Coal Association, 1935), p. 311 (mimeographed).

37. U. S. Bureau of Mines, as compiled in National Coal Association, *Bituminous Coal Annual, 1951* (Washington, 1952), p. 101. Secular changes in coal demand are discussed in Chap. 8.

38. U. S. Department of Commerce, *Survey of Current Business. National Income Supplement, 1951* (Washington, 1951), p. 207; and National Coal Association, *Bituminous Coal Data, 1951* (Washington, 1952), p. 2. The figures used to compute changes in gross national product are in current dollars.

Another aspect of demand accounts for the special concern of bituminous coal operators about downward price movements. Demand for coal, at least in the short run, tends to be inelastic with respect to price. Price reductions result, therefore, in lower total receipts, which explains why "price wars" are the bane of every coal operator's existence.

Prices and the Price Structure

Walton Hamilton has said of the pricing process in bituminous coal: "A lot of rumors about separate sales replace the market quotation; inadequate information, rarely verified, is circulated by written and oral report. Prices are for a commodity whose definition cannot be reduced to terms. There are no adequate criteria by which the terms of a contract may be made to reflect market conditions." [39]

Bituminous coal is not a homogeneous product. There are differences in quality, chemical content, volatility, friability (softness), and size. Coals are not interchangeable among different kinds of users. Certain coals well adapted for the production of steam in an electric utility, for example, are of no use in the manufacture of coke. Each type of coal, therefore, has its own conditions of demand and supply —its own selling price. There is no market price for bituminous coal; there is, rather, a complex price structure.[40] Because of differing conditions of demand and supply for the many kinds of coal, prices are constantly in flux. And relationships among the many prices tend to be unstable.

In the process of price making can be found the essential attributes of "pure" competition. There are many buyers and sellers in the typical market area. There is competitive bidding between buyers and sellers who are well informed about the quality of the product.[41] The notable imperfection is the incompleteness of information.

Few firms employ salesmen who call on prospective customers. Rather, the sales manager sits down each morning at the telephone with a list of buyers' names in front of him. Each consumer is quoted a price calculated to be low enough to attract custom but high enough

39. W. Hamilton, "Coal and the Economy—A Demurrer," *Yale Law Journal, 50* (February 1941), 598.

40. Illustrative of the elaborateness of the price structure is the minimum pricing schedule established in the early 1940's by the National Bituminous Coal Commission. The schedule included approximately 450,000 different selling prices.

41. No buyer, admittedly, can be sure that any given shipment from the seller will be as free of impurities as he desires. There is always the risk that the seller will allow his standards for cleaning of the coal to decline. For that reason, among others, buyers frequently make long-term contracts with sellers with a good reputation in the trade, even though such sellers may be charging slightly higher prices than their competitors. See below, p. 21.

to reward the seller. Very likely there will be bargaining between buyer and seller, the former perhaps remarking that a competitor of the latter has offered a somewhat lower price. The seller must decide whether to meet the lower quotation. He must consider questions of this kind:

1. Is this consumer bluffing?
2. Could the competing seller offer the lower price in good faith in view of what I know of his costs, cash asset position, and so on?
3. Will this customer, if offered the lower price, pass that information to other buyers and spoil my market?
4. In view of my costs, can I afford to cut my price?
5. If competitors are selling below me, can I afford to hold at my first asking price?

A decision will be made and the telephoning resumes. Thus is pricing information disseminated, prices of the competing sellers in each market tending toward equality.[42]

In many industries the seller sets his prices on a basis of unit cost plus a percentage markup. This rule of thumb tends to simplify pricing decisions. It cannot be applied, however, in bituminous coal mining, since coal is produced under conditions of joint cost. When the coal is blasted from the seam, it falls in a variety of sizes.[43] The operator has no alternative but to allocate costs to the different sizes by arbitrary decision. This means that he can only hope that his sales realization on total output will exceed his total producing costs.

Happily for the operator, domestic and commercial consumers prefer the larger sizes. Lump coal is sold at prices higher than the smaller sizes, partly because domestic and commercial users—who use comparatively small amounts of bituminous—have a bargaining position vis à vis producers which is considerably weaker than that of the great industrial consumers. The price differentials which arise are, of course, a reflection of the joint cost problem in mining. Each size has its own conditions of demand and supply, and as those conditions change there are adjustments in the relationships among the various prices.

It is to the interest of each mine manager to maintain a high level

42. Most firms in the southern Appalachian fields market their coal through a cooperative selling agency, Appalachian Coals, Inc. Information on prices is much more quickly exchanged among members than is common in the rest of the industry. See Chap. 2.

43. The use of carbon dioxide instead of powder for blasting permits somewhat greater control over sizes than in the past. But breakage during loading and sizing offsets, at least in part, the gains from using Airdox or Cardox. The producer must therefore count on getting sizes unwanted by any customer then on his books. This coal must be disposed of at what frequently amounts to "distress" prices.

of output. Each, therefore, will be anxious to engage a maximum number of buyers in long-term contracts (30 days or more).[44] For the consumer there is a similar benefit. He will want to do business with a firm which maintains high standards of quality, is prompt in deliveries, and takes care of its customers in the event of a sudden shortage of coal.

The contract itself is revealing of the unusual character of the coal business. It is full of loopholes for the producer. The seller is excused from shipping in the event of "strike, labor trouble, labor shortage, fires, flooding, or accident at the mine or mines where the coal sold hereunder is produced, embargoes, shortage of [railroad] car supply, contingencies of transportation, Federal, State or local laws or regulations, preference ratings established by any governmental authority, or other matters (whether of a like or dissimilar nature) beyond the control of the Seller which prevent or interrupt the shipment of such coal." [45] Only two provisions modify the position of the seller: 1. he undertakes to apportion coal equitably among contractees, and 2. he agrees that interpretations of "beyond the Seller's control" are to be made by the American Arbitration Association.

An escape clause is provided to allow for changes in selling price:

> The purchase price of coal specified herein is based, in part, upon production and sales cost, including the effect on such cost of Federal and State laws, including tax laws now in force. Any increase or decrease of such cost with respect to coal shipped hereunder caused by changes in the present agreements with mine employees or by the imposition of Federal, State or local authority of a direct charge or tax on coal, or on the sale or on the mining thereof, or on the payrolls or wages paid in the mining or sale or on transportation thereof, or by subsequent changes in the rate of such charge or tax . . . or by . . . legislation operating to limit the hours of labor or by any other laws or regulations . . . shall correspondingly increase or decrease said price of coal or any tonnage thereafter shipped hereunder.

About the only eventuality against which the seller is not protected is an increase in demand which results in rising prices. The buyer is the clear beneficiary in this case, if the contract extends for any considerable period after the price increase. On the other hand, the seller benefits when demand declines. It is not surprising, therefore,

44. Technically, coal is sold both in the "spot" market and under contract. For practical purposes nearly all coal is transferred under some sort of contractual arrangement, though some distress coal is exchanged by arm's-length bargaining on a day-to-day basis.

45. From a sample contract used by Wyatt, Inc., New Haven, Conn., a wholesale dealer in bituminous coal.

that when the coal is selling well the volume of long-term contracts increases. The reverse applies when demand for coal dips.[46]

Professor Hamilton has summarized well the pricing problems of the commercial bituminous operator:

To the operator the price is an act of finite judgement—he must make a bid which will get the order. The joint character of the product eliminates the significance of particular costs. A number of lots of different sizes have incurred a common cost of production; yet they must be sold at different prices. Since they take different freight rates as they move to different destinations, the net yield per ton shows a wide variation. The price an operator can afford to quote on a shipment depends upon his "realizations" from other shipments. It likewise turns upon the relative activity of his mine. Coal, like modern industry generally, is subject to diminishing [fixed] costs with increased volume. Consequently, every effort is made to maintain—and expand—sales in order to spread fixed costs thinly over as wide an area as possible. The concern of the seller is not with an item and its price, nor with a series of items each with its price. Instead, in an endless series of decisions concerned with a complex of interdependent prices, the operator must keep his business a going concern. ("Coal and the Economy," pp. 599–600)

The Impact of Costs on Market Position

The basis for differences in production costs among the competing regions has been established.[47] The differences have meaning only, however, when they affect (adversely or otherwise) the market position of the firms involved. If firms with relatively high costs must compete in the same markets with firms having lower costs, the former will be faced with selling problems.[48]

The pattern of distribution from mines in the Appalachian coal field is depicted in Table 3. The New England area is served primarily by District 1 (eastern and central Pennsylvania). The Middle Atlantic states are supplied by Districts 2 (western Pennsylvania) and 1, though the bulk of tonnage originates in the former. In the other four market areas most shipments come from District 8 (southwestern West Virginia and eastern Kentucky).

46. In the boom of 1920, for example, 90% of sales were under long-term contracts. See U. S. Coal Commission, *Report on the Effect of Irregular Operation,* Pt. 5, p. 3397. In the comparatively slow coal market of 1950, on the other hand, contract sales were estimated at 50% of all transactions, as is stated in a letter to the author from V. M. Johnston, Controller of Appalachian Coals, Inc., dated November 10, 1950.

47. See above, pp. 9–10.

48. Such is not always the case. Mines selling special-purpose coals can be less concerned about relative costs, especially if their customers cannot shift to competing fuels. In a real sense such mines are not competing with other commercial mines, for theirs is a differentiated product and they are selling in a different market.

Table 3. Percentage of Total Appalachian Area Sales, by Districts, to Various Markets, in 1946

	DISTRICT OF ORIGIN						
Region	*1*	*2*	*3*	*4*	*6*	*7*	*8*
New England	52%	7%	20%	—	—	12%	10%
Middle Atlantic	27	55	15	—	—	1	1
E. North Central	1	10	4	15	—	22	46
W. North Central	2	—	1	—	—	31	66
South Atlantic	5	7	11	—	—	28	47
E. South Central	—	—	—	—	—	7	93

Note: Totals may not add to 100% because of rounding.

Source: U. S. Bureau of Mines, *Bituminous Coal Distribution* (Washington, 1946), pp. 10–12.

The volume of sales by any district in a given market is conditioned principally by two things: selling price at the mine, which is influenced strongly by total production costs per ton; and transportation charges from mine to market. Both factors must be kept in mind, because a firm with relatively high production costs can still win custom if its delivered prices are lower than those of its competitors. That is the principal conclusion to be drawn from the data in Table 3.

As can be seen from that table, the tendency is for each market area to be dominated by that district which is geographically the nearest. District 1, despite its high production costs relative to Districts 2 and 3, outsells both by a wide margin in the New England area. Its advantage derives from lower transportation expense, which is upward of 50% of the delivered price of coal.[49] In the Middle Atlantic states, on the other hand, the advantage of eastern Pennsylvania in freight rates is more than offset by the lower production costs of western Pennsylvania. Mines in the latter district command the largest share of the market.

The impossibility of considering market position without taking into account both freight charges and production costs is demonstrated well in the east north central states, frequently referred to as the "lake trade." Nearly 70% of coal going to the lake ports in 1946 originated from the southern Districts 7 and 8. Average production costs in both districts were relatively high. And, of course, of all the Appalachian districts the southern fields are the most distant from the Great Lakes.

No single reason can be given for the predominance of southern sellers in the lake trade. Partly it is because freight rates from the

49. For a more thorough discussion of freight rates, see Chap. 2.

south were not proportional to the distance of the haul, and southern firms were able to offset their higher production costs by the favorable freight-rate differentials.[50] The large volume of sales from the south represents, in part, the desire of consumers in the lake region to obtain the "smokeless" coals which are peculiar to the southern Appalachian region. It partly reflects the existence of trade relations of long standing between southern producers and consumers in the lake region.

These few observations indicate that relatively high production costs do not automatically exclude a firm from all markets. The high-cost firms have been able to do business in markets where few of the firms in other districts can or desire to compete. Some high-cost mines, moreover, have been able to survive because of favorable freight-rate differentials.

Problems of Technological Change

Coal mining, it has been said, is a "discontinuous series of chores." Though machinery has been adopted to a considerable degree, it "remains that the most elaborate mechanization today is at best an extension of the 'materials-handling' philosophy, as if full-rigged ships were today equipped with nylon sails, aluminum hulls and plastic gear, but steam had not yet put in an appearance." [51]

The implied criticism is not entirely fair. While it is true that technological change in the coal industry amounts to the mechanization of processes formerly done by hand, there is no other way to mine coal. The alternatives, which were suggested by the same writer, would be gasification or hydrogenation of coal—which would yield in either case products quite different from bituminous coal. What is more, the delay in mechanizing the mines was in large measure a result of geological conditions. Even so, innovation—in the sense that new methods of organizing production are initiated—has proceeded steadily in recent years. Where cutting and loading were done by hand in the past, for instance, miners worked alone in their own rooms, setting their own pace. The introduction of loading machines necessitated a reorganization of mining methods to assure maximum utilization of the equipment. Each machine is assigned to a crew which moves from room to room under the supervision of a foreman. Similarly, the increasing speed of cutting and loading machines has required operators to install haulage equipment capable of handling greater loads.

The principal attraction of machinery to a coal producer is that it

50. The north-south freight-rate differential is discussed in Chap. 2.

51. "Coal," *Fortune, 35* (March and April 1947), 266.

tends to lower producing costs.[52] Mechanical devices effect cost reductions in any of three ways: 1. cut total outlays; 2. increase output per man per day (productivity); or 3. effect a combination of the two. If, then, unit production cost after mechanization is less, and if total profits (the difference between total receipts and total costs) are greater, there will be a motive for introducing new equipment into a mine.

The savings in production cost realized from mechanization must, of course, equal or, preferably, exceed the cost of maintaining and ultimately replacing the machines. The greater the excess of savings over maintenance and depreciation charges, the more willing is the operator to mechanize.

The most important consideration is the timing of technological change. The decision to modernize a mine is not simply one of determining whether there are cost savings to be realized. Because this matter is to be considered more thoroughly in a later chapter[53] the factors taken into account by the manager of the mine will simply be listed here with but brief comment. They include:

1. *The financial position of the firm.* Does it have adequate cash assets to purchase the machinery? If not, is it able to acquire credit in sufficient amount?

2. *The geological conditions in the mine.* Are the seams too thin to accommodate the newest equipment? Is the floor too scaly or does the roof need to be brushed in order to utilize the most efficient machines?

3. *The availability of equipment.* Are the machines currently being produced adaptable to the geological conditions of this mine?

4. *Present and anticipated demand for coal.* Do present and future prospects of total demand justify large investment outlays?

5. *Activities of competitors.* Are other firms mechanizing, thereby altering pre-existing competitive relationships with this firm? If so, will the market position of this firm be so changed as to demand that it follow the lead of the other firms?

6. *The level of costs in the firm.* Are costs now rising or are there indications that they will rise in the future to an extent that substitution of capital is a prerequisite for survival in the industry?

52. Some machinery, such as coal-washing equipment, is installed to enhance the marketability of the product. This is somewhat akin to the practice in the manufacturing industry of installing new machinery which will improve an existing product or produce a new one. There is evidence, in fact, that diversification of output is a principal reason for the purchase of new equipment in manufacturing firms. See, for example, L. G. Reynolds, "Toward a Short-run Theory of Wages," *American Economic Review, 38* (June 1948), p. 299.

53. See Chap. 7.

Past behavior indicates that innovations, once instituted, spread quickly throughout the industry, though not without some resistance. Some mine managers object to change as a violation of tradition. Others retain the older methods out of sheer inertia or unwillingness to reorganize their mines so as to accommodate the newer techniques.

There is evidence that these reluctant individuals ultimately mechanize if only because they must protect themselves against exclusion from the major markets. There is a tendency to follow the leader. A giant firm, such as Island Creek Coal Company, may place an order with a prominent mining-machinery producer. The lesser firms in the industry begin a scramble to maintain their competitive status. A sharp upswing in machinery sales results.[54]

Although some firms introduce machinery with reluctance, others initiate action. Their research and engineering departments are constantly seeking to improve productive efficiency by new methods and machines. The machinery manufacturer will be presented with a new idea, along with a request that the appropriate equipment be produced.[55]

The needs of buyers differ greatly depending on the physical conditions under which they operate. The machinery manufacturer must, therefore, produce a large number of different kinds of machines and an assortment of variations in each. Practically all equipment produced by the Joy Company is "tailor-made." In one type of loading machine, for example, the company has models with different loads, track gauges, voltages, gathering mechanisms, and heights of machine.

Inability to standardize equipment has an obviously adverse effect upon production costs and selling prices. Loading machines range from approximately $15,000 per unit to as high as $35,000. Shuttle cars are priced between $12,000 and $20,000 per unit. Cutting machines begin at $5,500 and run as high as $36,000. Joy's new continuous miner is priced at $56,000.[56] In an industry with as poor a profit record as bituminous coal, the surprising thing is that the industry has become as thoroughly mechanized as it has.

54. This appears to have happened in the mid-1930's. Sales engineers of the Joy Manufacturing Co. told me that the company, beginning in 1937, produced a whole new line of equipment on three different occasions for Island Creek. Data to be introduced in Chap. 7 indicate that it was just around that year that purchases of new loading equipment began to rise.

55. The Joy Co. reported to me that as much as 50% of its sales has been initiated by the purchaser. It should be noted, too, that research is being done by Bituminous Coal Research, a branch of the National Coal Association, and by the U. S. Bureau of Mines.

56. All prices quoted are as of June 1952.

SUMMARY

One of the distinguishing characteristics of the bituminous coal industry is the vigor of price competition among the many firms. The number of competing firms is large. Ease of entry into the bituminous industry permits a rapid increase in the business population on comparatively short notice. There is some concentration of output in the hands of about 300 firms out of a total of more than 6,000. And marketing agencies have sought to reduce interfirm price competition in the industry. Even so, there are several thousand medium and small-sized firms, producing 500,000 or less tons per year, which account for a combined tonnage of upward of 30% of the national output. These firms stand ever ready to reduce their selling prices should that be necessary for their continued existence in the industry.

The industry, then, is "workably competitive," a situation presumably in keeping with the public interest. At the same time there have been recurring complaints from within the industry that the rewards to the productive services have been inadequate—that returns to capital and labor have been less than in other occupations. This situation arises in large part from the fact that the demand for coal is highly elastic with respect to income. Small changes in the sales of the products of coal's major customers have a more than proportionate effect (in the same direction) on total consumption of coal. While there are many industries in a similar situation, the competitive structure of the coal industry enhances the possibility of competitive wage and price reductions during periods of declining demand. It is this vulnerability to downward price-wage spirals which has moved the industry to seek and the public to give assistance in the form of minimum pricing for both the product and the productive services.

Certain of the destabilizing influences in the industry have been eliminated or moderated in recent years. Union-enforced "floors" under wage rates have curbed the use of wage cutting as an instrument of competition, whether among firms or between regions. Moreover, labor has for a variety of reasons become increasingly mobile. As will be explained in a later chapter, these developments presently act as powerful deterrents to a recurrence of the troubles of the twenties.

CHAPTER 2

The Appalachian Region

INTRODUCTION

COAL DEPOSITS ARE to be found in nearly all sections of the United States.[1] The bituminous industry embraces all firms working the deposits, no matter where. It has been thought advisable here to study the effects of unionism on the major area of activity, the Appalachian region. Approximately 70% of the national coal output originates there. It is in the center of the principal market areas for bituminous. And it has been, particularly since 1933, the focal point of all wage bargains in the industry.

In the present chapter the region and its problems will be discussed. The first of three sections will be concerned with the identity and distinguishing characteristics of the component coal fields. In the second section market relationships of the competing regions, north and south, will be explored. In the third section the special position of the consumer-owned ("captive") mines relative to the commercial mines will be considered.

GEOGRAPHY AND GEOLOGY OF THE APPALACHIAN REGION

The Appalachian coal fields extend through eight states, though only six are coal-producing areas of importance. Ranked according to their total output in 1948, they are:[2]

1. West Virginia	168.8 million tons
2. Pennsylvania	134.5
3. Kentucky (eastern)	59.7
4. Ohio	38.7
5. Virginia	19.6
6. Alabama	18.0

1. The exception is New England and even there—in Rhode Island—coal was once mined. The quality of coal varies from section to section, ranging from the high-quality bituminous deposits in the Appalachians to the inferior lignite of the Dakotas.

2. U. S. Bureau of Mines, *Bituminous Coal . . . in 1948,* Mineral Market Survey No. 1807 (1949), p. 7. There are small deposits in Georgia, but total tonnage mined is not separately recorded.

7. Tennessee	6.5 million tons
8. Maryland	1.6

PRODUCTION DISTRICTS IN PRICE AREA I

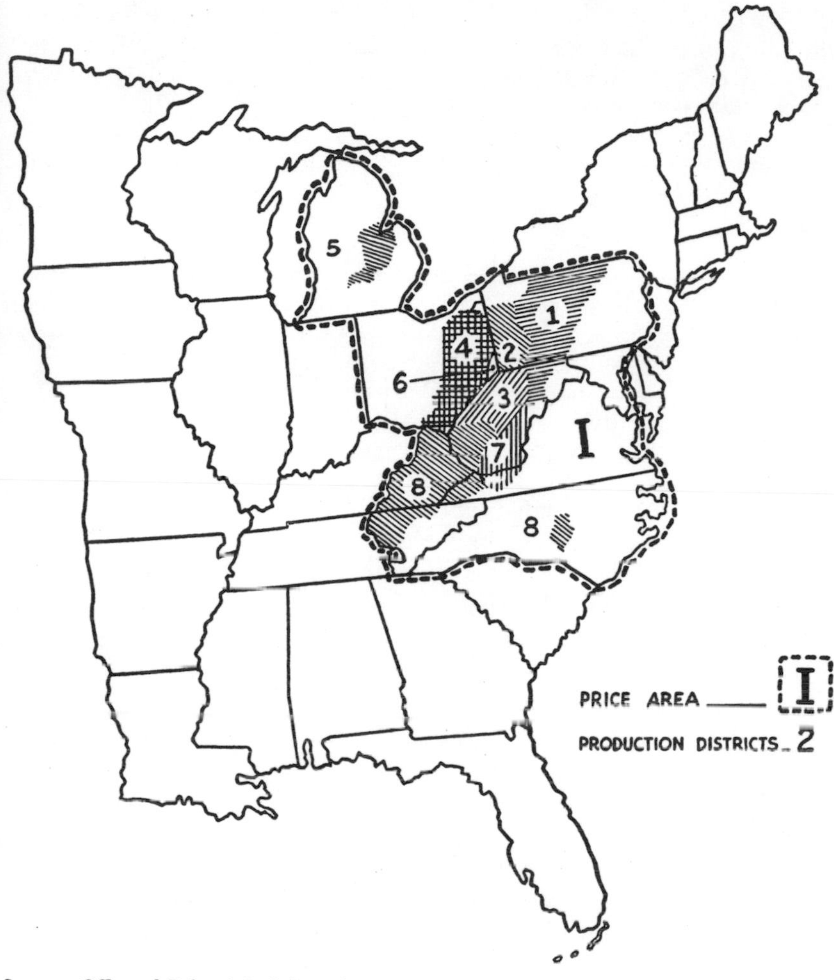

Source: Office of Price Administration

Although output in Alabama is substantial, relatively little of its tonnage competes with coal mined in the other states listed. For that reason Alabama is generally excluded from the Appalachian region.[3]

3. Comparatively little of Alabama's coal is mined for commercial sale, belonging as it does to industrial consumers. Under the National Bituminous Coal Commission and the wartime Office of Price Administration the individuality of Alabama was recognized by designating its coal fields as a separate pricing area.

For greater ease of governmental minimum and, later, maximum pricing programs, the bituminous industry was subdivided into numbered producing districts. Coal fields with substantially the same geological characteristics and in the same geographical locations were grouped together. Though the numbering system died with the dismantling of public price controls in 1946, much of the empirical data later to be introduced was collated under those designations. The areas were defined as follows:

DISTRICT 1: Central Pennsylvania, Maryland, and the Upper Potomac region
DISTRICT 2: Western Pennsylvania
DISTRICT 3: Northern West Virginia
DISTRICT 4: Ohio
DISTRICT 5: Michigan [4]
DISTRICT 6: Panhandle of West Virginia
DISTRICT 7: Southeastern West Virginia and western Virginia
DISTRICT 8: Southwestern West Virginia, eastern Kentucky, and northern Tennessee

Within the Appalachian area there are wide variations in the quality of coal, the amount of impurities in the seams, and the thickness of the veins. The richest fields tend to be concentrated in western Pennsylvania (District 1) and northern West Virginia (District 3). In general, the seams thin out, and impurities, such as drawslate, become more common as one moves in any direction away from those two areas.

The data in Table 4 highlight the advantage of Pennsylvania and West Virginia, where 26% and 28% respectively of coal output was extracted from mines of the "ideal" seam thickness, six to eight feet.

These geological differences account in large part for variations in production costs among the competing districts. As can be seen in Chart 1, total producing costs per ton were highest in central Pennsylvania, southern West Virginia and Virginia, and eastern Kentucky—producing Districts 1, 7 and 8. To some extent, however, the poorer physical conditions in certain areas are offset in the product markets by differences in quality. In Districts 1 and 7, for example, the seams are thin and producing costs are relatively high. But much

4. The output of coal fields in Michigan is very small. At the peak of coal demand during World War II, output amounted to but 140,000 tons of a total exceeding 600,000,000 tons. Very little data is available on costs, prices, and so on in the Michigan fields because separate tabulations would have revealed information about individual firms.

of the coal is of the semibituminous or smokeless kind, having a low ash and sulphur content. It is, as a result, an excellent steam coal and can find adequate markets for such use.[5]

Table 4. Per Cent of Bituminous Coal Produced in the Ten Largest Coal-producing States, by Thickness of Seam Mined, in 1945

State	*Under 4 Ft.*	*4 to 6 Ft.*	*6 to 8 Ft.*	*Over 8 Ft.*
Alabama	59	34	4	3
Virginia	49	40	2	9
Kentucky	42	47	11	0
Pennsylvania	36	32	26	6
West Virginia	32	34	28	6
Ohio	26	73	1	0

Source: U. S. Bureau of Mines.

Market Relationships of Competing Districts

Competition in the product markets must be among firms, no matter where located. Nonetheless, it is valid to note that mines in one region may have common interests vis à vis mines in another region. That has been true historically in the bituminous coal industry.

Bituminous mining in the southern Appalachians began to assume large proportions after 1900.[6] There were several reasons: the demand for coal was rising; there was excellent industrial steam coal in the south, particularly in West Virginia; mining was easier than in the north, because the mines were newer and the seams better; southern mines enjoyed preferential freight rates; and wage rates in the nonunion south were somewhat below those in the unionized north. These advantages combined to afford southern firms a distinct cost advantage over their northern competitors. Increasing quantities of southern coal flowed into markets previously served solely by mines in Pennsylvania, Ohio, Indiana, and Illinois (the so-called Central Competitive Field).

Attempts were made by the northern firms to redress the balance. Repeated pleas were addressed to the Interstate Commerce Commission for increases in the freight-rate differentials between north and

5. E. S. Moore, *Coal* (New York, John Wiley, 1922), pp. 374 f. See also A. M. Bateman, *Economic Mineral Deposits* (2d ed. New York, John Wiley, 1950), pp. 644–7.

6. It has been traditional to refer to mines south of the Ohio River as the "southern group"; those north of the river as the "northern group." As will be pointed out in Chap. 8, the distinction is now blurred, since northern West Virginia (District 3) is for practical purposes a part of the northern group.

south. The miners' union was pressed to organize the southern firms and impose upon them the same wage structure as that prevailing in the Central Competitive Field.

These efforts were only partially successful. The southern gains continued. The outbreak of World War I temporarily obscured the trend as total demand for coal reached hitherto unprecedented heights. Wartime controls on transportation facilities contributed to a disruption of "normal" channels of distribution, checking temporarily the increasing flow of southern tonnage into the major markets. The struggle between north and south was, therefore, postponed until the postwar period.

The end of World War I and the "reconversion" boom that followed signaled the resumption of interregional competition. None of the conditions which had made possible the striking prewar gains of southern operators had been altered. Two new factors entered the picture. First, the long-run upward trend in product demand came to a halt. High prices and recurrent shortages of bituminous during the war years stimulated among coal consumers much interest in greater economy in fuel use and in the potentialities of substitute fuels, i.e., petroleum and natural gas. Furthermore, technological changes —such as the increasing substitution of scrap for pig iron in steel fabrication—caused a reduction in the consumption of bituminous. After 1920, then, the commercial operators had to struggle among themselves for sales in markets where demand was constant, if not falling.

Second, the United Mine Workers at the war's end moved to make permanent the considerable gains it had registered during the "tight" labor market of the war years. It demanded that wartime earnings be maintained on the basis of a peacetime work day and work week. More important, it sought, with the blessings of the northern operators, to unionize the southern mines—thereby to protect the union wage scales in the north. Two particularly important work stoppages, in 1919 and in 1922, ultimately helped the union to solidify its position in the north. But from a long-range point of view the strikes were completely unsuccessful: the southern firms seized the opportunity to expand their sales while at the same time they were resisting effectively the attempt to unionize their mines.

These two factors, the end of the secular increase in coal demand and the union's inability to organize the south, were the underlying factors in the crisis which gripped the coal industry during the 1920's. Exploiting their cost advantages to the utmost, the southern firms steadily expanded their position in the major coal markets. Data on coal shipments from mines in Pennsylvania and Ohio on the one hand

and from West Virginia and Kentucky on the other depict dramatically the changes that were wrought. In 1920 shipments from the two northern states amounted to 60% of the total from the four states. By 1927 mines in Pennsylvania and Ohio were shipping only 39% of the total.

That union wage policy—in its broadest sense—was a basic influence in the convulsion cannot be doubted. Until 1927 the United Mine Workers stubbornly resisted efforts of the northern operators to lower the union scale. The hard-pressed northern producers were left with clear, though equally unpleasant, alternatives: either they abrogated their contracts with the union and operated on a nonunion basis or they awaited the inevitable financial disaster. Meanwhile, the steadfastness of the union was more than matched by the unyielding opposition to unionism among southern firms, significantly aided by the courts.

At length the untenable situation was resolved. When a strike in late 1927 failed to win for the union a renewal of the 1924 wage bargain with the northerners, the UMW reluctantly authorized its districts to make the best wage bargains possible. Wage rates all over the industry slumped toward the prevailing level in the south. And, as the data indicate, the decline in sales of northern mines, relative to those of the southern producers, came to an end. Northern firms, in fact, began in 1928 to recoup part of their earlier losses.

Freight rates were second only to wage rates as a weapon in interregional competition during the 1920's and 1930's. This was true for several reasons. Basically, it was because transportation charges amounted frequently to 40% or more of the delivered price of coal.

Freight rates are established nominally in terms of the cost of carriage. This means that, other things being equal, the rate tends to be proportionate to the distance of the haul. Were this actually the case, mines in certain parts of the south would be virtually excluded from the major markets in the industrial north and northeast. This obvious truth was recognized early by all parties in the bituminous industry.

The southern railroads were understandably anxious to encourage coal mining in the southern states. Firms in the south were equally interested in competing in the major coal markets. Track facilities were, therefore, extended into the southern coal fields. What is more, the carriers established rate schedules for southern coal which were low enough to permit the southern firms to compete freely in the northern markets. There was, that is, a "mutuality of interest" between carriers and southern operators. "Low freight rates [were] necessary to the operators, and if coupled with large tonnage, [were] possible for the railroads. Both parties [could] thrive on the combination and . . . united to maintain it against northern attacks." [7]

Table 5. Relationship of Mine Price and Freight Rates to Delivered Price per Ton of Bituminous

Year	*Average Price at Mine, per Ton*	*Average Freight Rate/Ton*	*Average Delivered Price/Ton*	*Per Cent of Mine Price to Delivered Price*
1928	$1.86	$2.27	$4.13	45
1929	1.78	2.25	4.03	44
1930	1.70	2.23	3.93	43
1931	1.54	2.22	3.76	41
1932	1.31	2.26	3.57	37
1933	1.34	2.20	3.54	38
1934	1.75	2.15	3.90	45

Source: Interstate Commerce Commission, "In Re: Increase in Freight Rates and Charges, 1935," Brief on Behalf of National Bituminous Coal Commission and Consumers' Counsel (1935), Pt. 1, p. 114.

When it became apparent to the northern firms that preferential freight rates were abetting the southern invasion of northern markets, they petitioned the Interstate Commerce Commission to increase the north-south differentials. In particular, northern operators were interested in the effects of the differentials on shipments to ports on the Great Lakes. "The controversy . . . struck to the heart of the operators' problem of regular production, because the lake trade acts as a balance wheel for those operators who secure it. By providing a tonnage demand in midsummer when industrial and domestic consumption is seasonally lowest, it reduces the wastes of irregularity." [8]

There is no need here to recount the details of a struggle which has been well described by H. C. Mansfield and others. The immediate issue throughout the long debate between the regions was the amount of the rate differential between northern and southern shippers. A variety of pressures, economic and political, were brought to bear on the ICC. The Commission reacted to these pressures as might be expected; it wavered, took a stand, retreated, took a new stand. Whatever position it adopted at a given time, its decision directly affected market relationships in the industry. Without question, the marked

7. H. C. Mansfield, *The Lake Cargo Rate Controversy* (New York, Columbia University Press, 1932), p. 39.
8. Ibid., p. 30.

gains of the southern coal operators throughout the early 1920's can be traced in large measure to the success of the southern operators and their carriers in preventing the ICC from altering significantly the freight-rate differential between north and south.

The restoration of a "balance of power" in the Appalachian region after 1927 did not terminate the north-south struggle for markets. Competition became more intense as the full effects of the cyclical decline after 1929 came to be felt. As total demand for coal declined, prices and wage rates were repeatedly slashed throughout the industry. Pleas arose in many quarters that the industry be "rationalized"—that plans be devised and implemented under which the number of competing firms would be diminished and selling prices shored up. But the bitter divisions between and within the competing regions precluded such arrangements, whether under public or private auspices.

At length, in late 1931, a committee of 50 coal operators representing the National Coal Association convened for the purpose of establishing regional sales committees. These committees were to sell jointly the output of all members, effecting a reduction in selling costs for each firm. Mergers of competing firms were to be encouraged. More effective merchandizing of coal to combat the competing fuels was to be carried on.[9] Though it was never stated so baldly, the committees were to attempt through cooperative selling to set and maintain prices at the highest possible levels.

The 50 coal producers invited to the meeting represented "the bulk of the commercial tonnage east of the Mississippi River." [10] But when the conference was assembled, there were 200 producers in attendance—all from the southern mines in Kentucky, West Virginia, Virginia, and Tennessee. This group organized itself into Appalachian Coals, Inc., with the promise that "similar corporations are to be organized for other districts." [11] The organizers of ACI probably had a motive more concrete than simple monopolistic price fixing. "For the specific group which organized it, the chief operators of the southern high volatile field, the agency . . . was a weapon of great promise in their struggle to reverse the trend against them in the interregional competition for coal business. It offered the southern fields an excellent chance to regain business which had been lost to Ohio and Pennsylvania after 1927." [12]

9. *New York Times,* December 5, 1931, p. 6.
10. Ibid., p. 6.
11. *New York Times,* January 7, 1932, p. 39.
12. E. V. Rostow, "Bituminous Coal and the Public Interest," *Yale Law Journal,* 50 (1941), 543, 558.

It may appear at first glance that price fixing by a relatively small segment of the industry was an unlikely method of recouping lost markets. Attempts to raise prices above the prevailing level would, on the contrary, tend to enhance the competitive position of the northern firms. Moreover, dissenters from the Rostow theory have pointed out that northern operators were invited to join ACI or to create their own selling agencies on similar lines. For their part the southerners professed publicly their interest in increasing coal consumption in competition with petroleum, natural gas, and electric power.[13] No indications were given that ACI was to be deployed in the battle for markets.

In the light of contemporary events, however, the Rostow view seems sound. The northern fields after 1927 had begun to recoup their earlier losses to southern firms. The bitter competition in the lake trade was at its height, and even as ACI was being organized the north was bemoaning a new setback in its efforts to get the Interstate Commerce Commission to increase the freight differential on lake shipments.[14] Animosity between the two regions was never greater.

All parties must have recognized, as did the Supreme Court in 1933, that ACI could have little effect by itself on the bituminous price level.[15] Its most effective program would have to be merchandizing. This could only mean that southern output would be promoted vigorously as against competing coals. In its arguments before the Supreme Court the agency made this quite clear: its purpose was to "enable the producers in this region [the south], through the larger and more economic facilities of such selling agency, more equally to compete in the general markets for a fair share of the available coal business." [16]

The northern operators must have recognized that the primary purpose of ACI was to advance the interests of southern operators at their expense. They declined the invitation to join the agency, despite their interest in any plan which would raise prices while not otherwise restricting their freedom of action.[17] In defense of their refusal to join, the northern firms pointed out that the legality of joint

13. *New York Times,* January 7, 1932, p. 39. See also Hamilton, "Coal and the Economy," *Yale Law Journal, 50* (February 1941), 605.

14. *New York Times,* February 14, 1932, Sec. 3, p. 6.

15. Appalachian Coals v. U.S., *U. S. Reports, 288* (1933), 344.

16. Ibid., p. 366.

17. The northern producers strongly opposed the Davis-Kelly Bill of 1932, the direct ancestor of the National Industrial Recovery Act of 1933. Besides their objection to governmental supervision of the industry, they abhorred the UMW-sponsored provision in the Davis-Kelly Bill for a guarantee of collective bargaining. *New York Times,* February 14, 1932, Sec. 3, p. 6.

selling agencies was in doubt and that they would await a decision in the expected test case.[18]

The legality of Appalachian Coals, Inc. having been affirmed by the Supreme Court, the agency has remained active in the industry. Its membership is still drawn almost totally from the southern fields. It is an ardent spokesman for the southern interests, working in close harmony with the Southern Coal Producers' Association. This latter relationship is in itself significant, since the SCPA has persistently fought to "protect" the south against the "depredations" of what appears to them to be a coalition between the United Mine Workers and the northern operators.[19]

The Special Position of the Captive Mines

While not all large consumers of bituminous coal own and operate mines, the motives for doing so are evident. The user is protected against inadequacies of supply and inordinate price increases brought on by sudden increases in demand. While this is an important reason for operating captive mines, the owner must still face up to many of the problems encountered by commercial operators: work stoppages, upward pressure on wage rates, temporary shortages of transportation facilities (if the captive mine depends on the common carriers), and so on. Consumer operation of a coal mine is advisable only if the mine can be operated at least as efficiently (in terms of cost) as the least efficient commercial firm which would supply the needs of this particular consumer. Put a bit more precisely, the long-run average costs of the captive mine must more often be below than above expected market prices over a period of time. Otherwise the captive operator would be exposed to the risk that market prices would fall below the price he can justifiably charge for his own coal.

The chief concern with captive mines here is their relationship to the commercial firms. For one thing, the captives reduce the potential sales of the commercial operators. Consumer-owned mines produce about 17%—upward of 100,000,000 tons—of the annual national output.[20] A substantial volume of sales is thus denied to the independent sellers.

In another respect, too, the behavior of the captives intimately affects the commercial operators. In periods when the steel industry, for example, is anxious to maintain high levels of steel output, managers will exert every effort to avoid work stoppages arising out of

18. Ibid., p. 6.

19. North-south relations since 1933 are discussed in detail in later chapters, especially Chap. 7.

20. In 1948 captive mines produced 95,000,000 tons of a total output of 599,500,000 tons. National Coal Association, *Bituminous Coal Annual, 1951*, pp. 50, 68.

labor disputes in their coal mines. Their resistance to union demands will be lessened appreciably. At the same time the commercial operators may be faced with shrinking demand and sizable stockpiles of unsold coal. Their interest will be surely in resisting the union, even if a protracted work stoppage should ensue.

All too frequently in recent years just such a situation has obtained. Activity in the steel industry, the electric utilities, and the railroads has been at continuously high levels, while the bituminous industry has languished in semidepression for much of the time. Several times since World War II the union has succeeded in splitting the captives away from the rest of the bituminous industry, has achieved much of its demands, and has then imposed the settlement on the commercial operators. That this procedure has stirred the helpless wrath of the latter is well indicated by the angry statement that "the steel industry or someone else is going to write the contract." [21]

SUMMARY

The Appalachian coal region is highly diverse. The biuminous operators seem to have little in common, much at variance. Geological conditions dictate that their producing costs will be different. Until recently their wage bills were different. Their interests are constantly clashing. Only their markets are the same—and "there's the rub."

The problems confronting the bituminous producers would be troublesome even if the demand for coal were rising. That is the lesson of the history of the industry prior to World War I. Since 1920 bituminous coal has barely held its own in absolute terms in the fuel markets. Relative to the substitute fuels it has been losing ground steadily.

The interesting question with which this study is principally concerned is, how has the union affected the complex problems of the industry? Has it brought increased stability by imposing its own solutions on the problems? Or has the UMW, pursuing short-run objectives, created in the long run more serious difficulties for the industry?

21. See, for example, the *New York Times*, May 26, 1949, p. 18. The issue of union tactics in wage bargaining is discussed in detail in Chap. 5.

CHAPTER 3

Significant Developments in the Industry, 1890–1950

INTRODUCTION

UNION WAGE POLICY cannot be made in a vacuum. While, as will be seen in later chapters, the activities of the United Mine Workers have had important effects on the bituminous industry, the union has had at times to adapt its short-run policies to the prevailing conditions. For this reason it is necessary to review the trends of key variables in the industry from the date of the organization of the UMW in 1890 to 1950.

From 1890 to 1919 a secular increase in coal demand sparked a steady expansion of the industry. Basic problems in relationships among the firms were thereby obscured. After 1920 and until 1929, however, the demand for bituminous flattened out. Interregional competition, which had been growing in intensity throughout the prewar period, reached a crescendo in the late twenties. Thereupon the industry was struck by the hammer blows of the 1930–32 depression in general business activity. With the advent of the New Deal important changes were wrought. Demand for coal began to ascend. Prices, profits, and production increased, reaching 20-year peaks during World War II. Most important, the years 1933–50 marked a spectacular growth in the power of the miners' union.

The raw material of this chapter is statistical data on prices, output, employment, mine population, unit costs, and wage rates. To minimize references in the text to specific figures, a composite table and accompanying charts have been inserted for reference.

TRENDS, 1890–1919

Output of coal increased steadily once the depression of 1893–96 had ended. Tonnage mined in 1900 was nearly double that of 1890 and that of 1910 was nearly double the production of 1900. Between 1890 and 1918 annual tonnage increased fivefold, an impressive figure, although 1918 was a year of "abnormal" demand.

Table 6. Growth of the Bituminous Coal-mining Industry in the United States, 1900–51

Year	*Production (net tons)*	*Total* *	*Average per Ton* *	*Men Employed*	*Number of Mines*
1900	212,316,112	$ 220,930,313	$1.04	304,375	—
1901	225,828,149	236,422,049	1.05	340,235	—
1902	260,216,844	290,858,483	1.12	370,056	—
1903	282,749,348	351,687,933	1.24	415,777	—
1904	278,659,689	305,397,001	1.10	437,832	4,650
1905	315,062,785	334,658,294	1.06	460,629	5,060
1906	342,874,867	381,162,115	1.11	478,425	4,430
1907	394,759,112	451,214,842	1.14	513,258	4,550
1908	332,573,944	374,135,268	1.12	516,264	4,730
1909	379,744,257	405,486,777	1.07	543,152	5,775
1910	417,111,142	469,281,719	1.12	555,533	5,818
1911	405,907,059	451,375,819	1.11	549,775	5,887
1912	450,104,982	517,983,445	1.15	548,632	5,747
1913	478,435,297	565,234,952	1.18	571,882	5,776
1914	422,703,970	493,309,244	1.17	583,506	5,592
1915	442,624,426	502,037,688	1.13	557,456	5,502
1916	502,519,682	665,116,077	1.32	561,102	5,726
1917	551,790,563	1,249,272,837	2.26	603,143	6,939
1918	579,385,820	1,491,809,940	2.58	615,305	8,319
1919	465,860,058	1,160,616,013	2.49	621,998	8,994
1920	568,666,683	2,129,933,000	3.75	639,547	8,921
1921	415,921,950	1,199,983,600	2.89	663,754	8,038
1922	422,268,099	1,274,820,000	3.02	687,958	9,299
1923	564,564,662	1,514,621,000	2.68	704,793	9,331
1924	483,686,538	1,062,626,000	2.20	619,604	7,586
1925	520,052,741	1,060,402,000	2.04	588,493	7,144
1926	573,366,985	1,183,412,000	2.06	593,647	7,177
1927	517,763,352	1,029,657,000	1.99	593,918	7,011
1928	500,744,970	933,774,000	1.86	522,150	6,450
1929	534,988,593	952,781,000	1.78	502,993	6,057
1930	467,526,299	795,483,000	1.70	493,202	5,891
1931	382,089,396	588,895,000	1.54	450,213	5,642
1932	309,709,872	406,677,000	1.31	406,380	5,427
1933	333,630,533	445,788,000	1.34	418,703	5,555
1934	359,368,022	628,383,000	1.75	458,011	6,258

* Figures for the years 1900 to 1936, inclusive, and 1939, exclude selling expense.

Year	Production (net tons)	Total	Average per Ton	Men Employed	Number of Mines
1935	372,373,122	658,063,000	1.77	462,403	6,315
1936	439,087,903	770,955,000	1.76	477,204	6,875
1937	445,531,449	864,042,000	1.94	491,864	6,548
1938	348,544,764	678,653,000	1.95	441,333	5,777
1939	394,855,325	728,348,366	1.84	421,788	5,820
1940	460,771,500	879,327,227	1.91	439,075	6,324
1941	514,149,245	1,125,362,836	2.19	456,981	6,822
1942	582,692,937	1,373,990,608	2.36	461,991	6,972
1943	590,177,069	1,584,644,477	2.69	416,007	6,620
1944	619,576,240	1,810,900,542	2.92	393,347	6,928
1945	577,617,327	1,768,204,320	3.06	383,100	7,033
1946	533,922,068	1,835,539,476	3.44	396,434	7,333
1947	630,623,722	2,622,634,946	4.16	419,182	8,700
1948	599,518,229	2,993,267,021	4.99	441,631	9,079
1949	437,868,036	2,136,870,571	4.88	433,698	8,559
1950	516,311,053	2,500,373,779	4.84	415,582	9,429
1951	535,000,000	2,605,450,000	4.87	400,000	9,300

		NET TONS PER MAN		PER CENT OF UNDERGROUND PRODUCTION		PER CENT OF TOTAL PRODUCTION	
Year	Average Number of Days Worked	Per Day	Per Year	Cut by Machines *	Mechanically Loaded	Mechanically Cleaned †	Mined by Stripping
1900	234	2.98	697	24.9	—	—	—
1901	225	2.94	664	25.6	—	—	—
1902	230	3.06	703	26.8	—	—	—
1903	225	3.02	680	27.6	—	—	—
1904	202	3.15	637	28.2	—	—	—
1905	211	3.24	684	32.8	—	—	—
1906	213	3.36	717	34.7	—	2.7	—
1907	234	3.29	769	35.1	—	2.9	—
1908	193	3.34	644	37.0	—	3.6	—
1909	209	3.34	699	37.5	—	3.8	—

* Percentages for the years 1900 to 1913, inclusive, are of total production, as a separation of strip-mine and underground production is not available for these years.

† For the years 1906 to 1926, these percentages are exclusive of coal cleaned at central washeries operated by consumers; after 1926, when data became available on the tonnage cleaned by consumer-operated plants, the percentages include the total tons cleaned at the mines and at consumer-operated washeries.

Table 6 (continued). Growth of the Bituminous Coal-mining Industry in the United States, 1900–51

		NET TONS PER MAN		PER CENT OF UNDERGROUND PRODUCTION		PER CENT OF TOTAL PRODUCTION	
Year	*Average Number of Days Worked*	*Per Day*	*Per Year*	*Cut by Machines*	*Mechanically Loaded*	*Mechanically Cleaned*	*Mined by Stripping*
1910	217	3.46	751	41.7	—	3.8	—
1911	211	3.50	738	43.9	—	—	—
1912	223	3.68	820	46.8	—	3.9	—
1913	232	3.61	837	50.7	—	4.6	—
1914	195	3.71	724	51.8	—	4.8	0.3
1915	203	3.91	794	55.3	—	4.7	.6
1916	230	3.90	896	56.9	—	4.6	.8
1917	243	3.77	915	56.1	—	4.6	1.0
1918	249	3.78	942	56.7	—	3.8	1.4
1919	195	3.84	749	60.0	—	3.6	1.2
1920	220	4.00	881	60.7	—	3.3	1.5
1921	149	4.20	627	66.4	—	3.4	1.2
1922	142	4.28	609	64.8	—	—	2.4
1923	179	4.47	801	68.3	0.3	3.8	2.1
1924	171	4.56	781	71.5	.7	—	2.8
1925	195	4.52	884	72.9	1.2	—	3.2
1926	215	4.50	966	73.8	1.9	—	3.0
1927	191	4.55	872	74.9	3.3	5.3	3.6
1928	203	4.73	959	76.9	4.5	5.7	4.0
1929	219	4.85	1,064	78.4	7.4	6.9	3.8
1930	187	5.06	948	81.0	10.5	8.3	4.3
1931	160	5.30	849	83.2	13.1	9.5	5.0
1932	146	5.22	762	84.1	12.3	9.8	6.3
1933	167	4.78	797	84.7	12.0	10.4	5.5
1934	178	4.40	785	84.1	12.2	11.1	5.8
1935	179	4.50	805	84.2	13.5	12.2	6.4
1936	199	4.62	920	84.8	16.3	13.9	6.4
1937	193	4.69	906	—	20.2	14.6	7.1
1938	162	4.89	790	87.5	26.7	18.2	8.7
1939	178	5.25	936	87.9	31.0	20.1	9.6
1940	202	5.19	1,049	88.4	35.4	22.2	9.4
1941	216	5.20	1,125	89.0	40.7	22.9	10.7

		NET TONS PER MAN		PER CENT OF UNDERGROUND PRODUCTION		PER CENT OF TOTAL PRODUCTION	
Year	*Average Number of Days Worked*	*Per Day*	*Per Year*	*Cut by Machines*	*Mechanically Loaded*	*Mechanically Cleaned*	*Mined by Stripping*
1942	246	5.12	1,261	89.7	45.2	24.4	11.5
1943	264	5.38	1,419	90.3	48.9	24.7	13.5
1944	278	5.67	1,575	90.5	52.9	25.6	16.3
1945	261	5.78	1,508	90.8	56.1	25.6	19.0
1946	214	6.30	1,347	90.8	58.4	26.0	21.1
1947	234	6.42	1,504	90.0	60.7	27.7	22.1
1948	217	6.26	1,358	90.7	64.3	30.2	23.3
1949	157	6.43	1,010	91.4	67.0	35.1	24.2
1950	183	6.77	1,239	92.6	69.4	38.5	23.9
1951	191	7.00	1,338	—	71.0	42.1	22.1

Source: U. S. Bureau of Mines

Prices, too, experienced a long-run upward movement between 1890 and 1920, but they moved in a more narrow range than did output.[1] In the first ten years of the period, prices declined approximately 12%, while output was doubling. Between 1900 and the outbreak of World War I average value rose 13¢ per ton, an increase of 13%. During the same period total tonnage mined expanded from 212,000,000 tons to 422,000,000 tons, a gain of nearly 100%. Not until 1917, the beginning of the wartime industrial expansion in the United States, did coal prices move up sharply. From a level of $1.32 per ton in 1916 average mine realization hit a peak of $3.75 per ton in 1920. This was a rise of 184%.

For most of 1890–1920, then, output increased tremendously while prices edged up rather slowly. This indicates that rising demand for bituminous coal was being met by increasing supply. The data in Table 6 on the number of men employed and on the number of active mines confirm this conclusion.

The number of men employed in the industry increased from less than 200,000 in 1890 to more than 600,000 in 1920. Moreover, the increase in total men employed was virtually continuous—in only four years during the 30 covered were there declines in total employment.

1. The data given are not market prices, but represent instead the average mine realization per ton of coal sold. The latter figure is obtained by dividing total tonnage sold by total dollars received. The quotient is not the same as market price, because coal is sold at any given time at a variety of prices.

CHART 3

Trends in Key Variables, 1890–1950

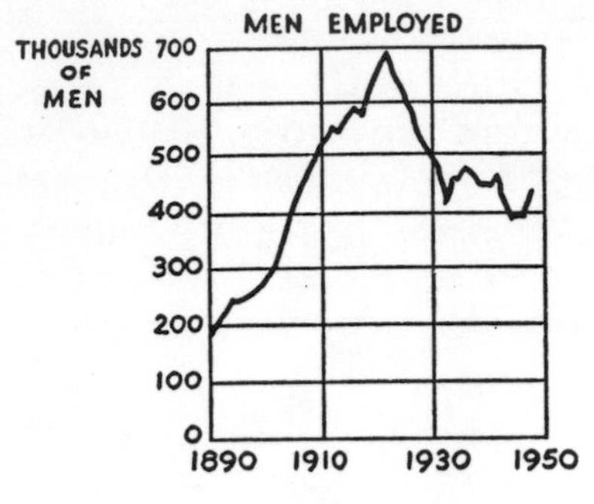

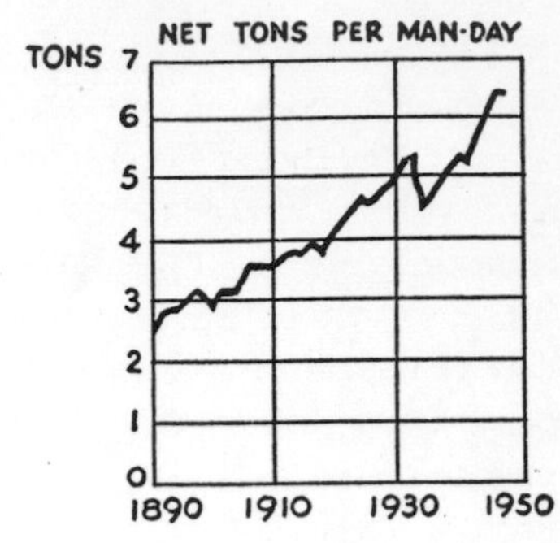

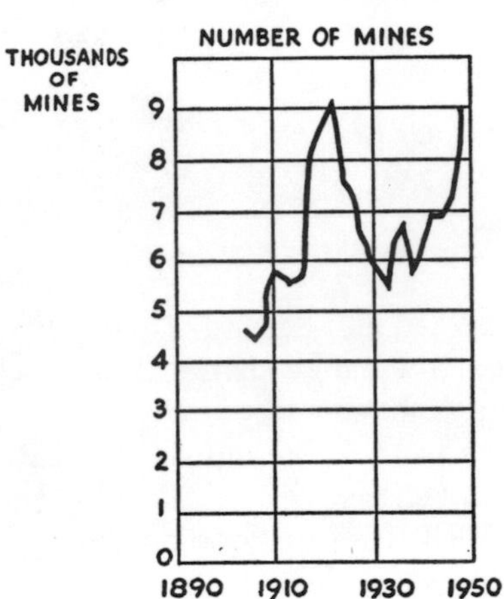

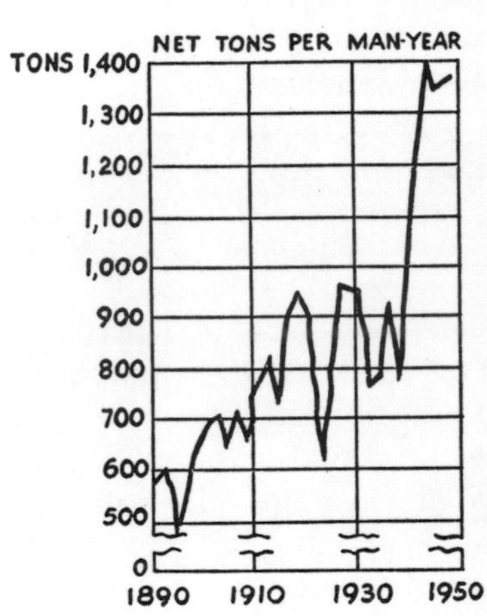

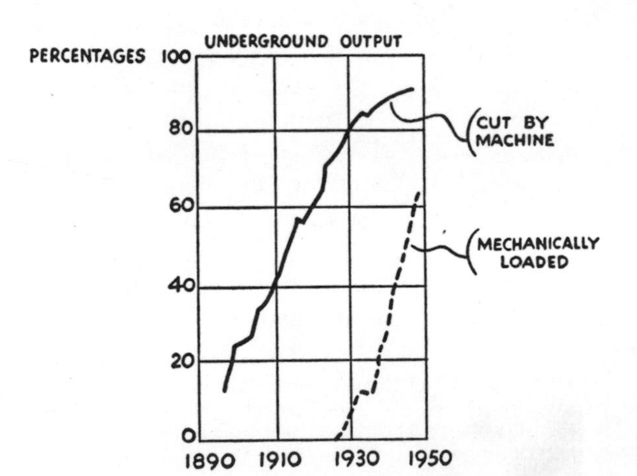

Source: See Table 6

These losses were quickly recouped and the upward trend resumed in each case.

To a substantial degree the increased employment was the result of an expansion in the number of operating mines in the industry. Mine population rose from 2,500 in 1895 to 5,600 in 1914 and reached a maximum of 9,000 in 1919. In addition, it is quite likely that existing mines expanded as demand for coal rose, accounting for part of the increase in total employment. The great increase in coal output, then, can be traced in large measure to the expansion in both total men employed and total operating mines. In addition, the data in Table 6 indicate that productivity was rising steadily, primarily because of technological improvements in the process of mining. The principal factor was the quickened pace of introduction of cutting machines into the bituminous mines.[2]

Though the relationship between increased productivity and mechanization needs little discussion, it is well to point out that as the percentage of output cut by machine rose from 5% in 1890 to 60% in 1920, output per man per day—a rough measure of productivity—rose 56%.

It has been estimated that labor cost per ton in bituminous mining was, prior to World War I, in excess of 70% of average total cost.[3] Under such conditions it is evident that a change in labor cost would have a direct impact on total cost. Furthermore, roughly 60% of the labor force in the industry was being paid on a piece-rate basis.[4] Changes in wage rates, then, were very likely to have important effects on unit labor costs.

On the basis of very scanty evidence it appears that labor costs per ton remained relatively stable from 1890 to 1916, after which greatly increased product demand brought about an upward shift in the demand for coal labor and rising wage rates. One reason for the relative stability of wage rates prior to World War I was the inability of the United Mine Workers to organize all of the competing coal fields. Perhaps more important, wage rates in alternative occupations were relatively unchanged during this period.

It has been pointed out that in periods of declining product demand coal operators had sought to remedy declining profits by reducing

2. To simplify the blasting operation and particularly to reduce pulverization of coal in the seam, the face is kerfed or undercut. Prior to the development of the cutting machines, this work was done by a miner who painfully picked out the kerf, often from a prone position. The introduction of a machine, of course, speeds the whole operation. It also eliminates one of the most uncomfortable and hazardous tasks in the mine.

3. W. Fisher and A. Bezanson, *Wage Rates and Working Time in the Bituminous Coal Industry, 1912–1922* (Philadelphia, University of Pennsylvania Press, 1932), p. 5.

4. Ibid., p. 66.

wage rates, the largest single item in their budgets. To protect themselves against wage cuts the miners succeeded in 1898 in establishing more or less uniform wage rates in the coal fields of Illinois, Indiana, Ohio, and western Pennsylvania. The union was unable, however, to secure a similar agreement with operators in the expanding coal fields of West Virginia, Tennessee, and Kentucky. Producers in the latter regions were able to pay lower wages than their northern competitors and to market increasing quantities of coal in consuming areas which were formerly the exclusive preserve of the northern operators. Rather than being able to exploit their newly won recognition from the operators, the miners were constantly fighting to preserve what gains they had made previously. For the industry as a whole, therefore, money wage rates rose only slowly, as the derived demand for labor increased in response to rising demand for the product. The slow rate of increase of money wage rates, coupled with rising productivity of labor, caused a leveling out of the movement of labor cost per ton. There were no appreciable increases in unit labor costs betwen 1903 and 1915, after which both wage rates and labor costs per ton began a sharp ascent.[5]

The period 1890 to 1920 was one of growth in the bituminous coal industry. The volume of output increased fivefold. Employment tripled. The number of active mines more than tripled. Output per man-day rose from 2.5 to 4 tons. Coal mined by machine rose from 5% of total output to more than 60%, and the efficiency of cutting machines was steadily improved.

Within the industry, however, important adjustments were under way. Now and then, when the demand for coal became temporarily depressed, price wars broke out and the industry became gripped in a downward price-wage spiral. To prevent such outbreaks of cutthroat competition a large segment of the industry accepted the principle of collective bargaining and the establishment of minimum wage rates for its employees.

This tactic was only partially successful, however, because only part of the industry subscribed to the wage agreement. Nonunion areas, particularly in the southern Appalachian fields, firmly resisted unionization. With excellent geological conditions for mining, and enjoying a preferential freight rate and wage rate, the southern fields began a steady encroachment on markets formerly dominated by northern producers. The outbreak of war in 1917 temporarily checked the decline in sales of the northern firms, relative to the southern. War and postwar demand for coal pushed prices to record heights, and all producers were able to sell whatever they could mine. Though producing costs

5. See Table 3-1 in appendix.

rose, too, they lagged behind selling prices, so profits for nearly all coal operators between 1917 and 1920 were substantial. At the close of the first World War the bituminous industry looked robust indeed.

Trends, 1920–29

The situation in the bituminous industry became increasingly unpleasant after the early 1920's. Within a few years after the war's end what had been a vigorous and expanding industry was being characterized as sick. Congress launched an intensive study of the industry, creating a special commission to conduct the inquiry. Responsible union officials became panicky, too; as early as 1922 they were openly advocating nationalization of the coal mines.

The changes that took place are mirrored in the movements of the key variables. Average mine realization fell from $3.75 in 1920 to $1.78 in 1929. Output, on the other hand, remained virtually unchanged. But there was a decline of 136,000 or 21% in the number of men employed. Wage rates dropped from an average of $7.50 per day in the northern mines to an average of $5.50. Where in 1920 more than 90% of bituminous coal corporations reporting to the Bureau of Internal Revenue enjoyed profits, by 1929 only 35% were in that position.[6]

The central factor in the situation was the leveling off of the demand for coal. Two reasons accounted for the ending of the secular increase in demand: high interest of coal consumers in economy in fuel consumption, and a trend toward the adoption of competitive fuels. Both changes were a direct result of the unusually high prices of bituminous coal during the war years. Many consumers, moreover, were plagued by intermittent interruptions in their fuel supply, interruptions caused by work stoppages, shortages of transportation facilities, and so on. These consumers shifted to substitute fuels, in part to free themselves from dependence on a single source of energy. Thus the great expansion in American productive capacity between 1920 and 1929 was powered by virtually the same quantity of coal as had been used at the start of the period.

The secular decline in demand for bituminous compared to competing fuels heightened the effects of interregional competition. During the years before the war rising coal demand had enabled northern firms to increase their sales in most markets in absolute terms, though their relative share of the trade in some areas had declined. With product demand at a plateau after the war, the gains of the south at the expense of the north were both relative and absolute. Chapters elsewhere in

6. Data from U. S. Bureau of Internal Revenue, as compiled in National Coal Association, *Bituminous Coal Data, 1935–1948* (Washington, 1949), p. 94.

this study contain descriptions of the effects of this postwar situation, particularly as it affected the northern firms and the United Mine Workers.[7]

Changes, 1929–32

The bituminous industry, badly weakened by the conflicts of the twenties, was hit severely after 1929 as the demand for coal receded during the general cyclical downturn. A few statistics will help measure the effects of the depression of the industry: output down 42%; average mine realization down 26%; number of men employed down 19%; number of active mines down 10%.

No region and relatively few firms managed to escape the disaster. By 1932 only 16% of 1,900 reporting firms had a net income after taxes.[8] So hard was the industry hit that prices, wage rates, output, and profits diminished to levels not touched since the turn of the century.

Trends, 1933–50

The story of the bituminous industry between 1933 and 1950 is dominated by what economists call "exogenous" forces—factors external to the industry. From 1933 to 1946 the bituminous industry was almost continuously under some form of supervision by the Federal Government. And from 1946 to 1949 the government was a party, willing or not, to every major dispute in the industry. Most important, the Federal Government after 1933 made possible not only the revivification of the United Mine Workers, but also the establishment of that union as the most powerful single labor organization in the United States.

Under the stimulus of the recovery measures instituted by the New Deal, the bituminous coal industry began to revive in 1933. Excepting the general setback in the economy in 1938, coal consumption rose steadily. From a low of slightly more than 300,000,000 tons in 1932, output reached 531,000,000 tons in 1948. Relative to substitute fuels, however, demand for bituminous continued to decline.[9] As consumption increased, so did average mine realization, wage rates, and employment. The number of active mines rose. The rate of introduction of mining machinery was substantially increased. In short, as business in general revived, the demand for coal increased, and prosperity was restored slowly to the coal industry.

This period, especially from 1933 to 1941, was one of social and

7. See Chaps. 2 and 4.
8. Data from U. S. Bureau of Internal Revenue, as compiled in National Coal Association, *Bituminous Coal Data, 1935–1948,* p. 94.
9. See Table 3-2 in appendix.

economic experimentation. As a striking example of a "sick" industry, bituminous coal was an early and continuing target for public assistance. Under the NRA one of the first codes to be rushed through was that for bituminous coal. The tacit encouragement of the President of the United States appeared to be behind the United Mine Workers when in 1933 the union sent into the coal fields hordes of organizers shouting, "The President of the United States wants you to join the union!"

When the NRA was declared unconstitutional in 1935, a "little NRA" for bituminous coal was rushed through Congress reportedly with a suggestion from the President that the legislators leave any worries about the law's constitutionality to the courts.[10] And when the first Guffey Coal Act was itself declared unconstitutional, a second was quickly passed by the Congress. At every turn, it appeared, a solicitous government stood ready to bail out the industry.

The details of the laws, all of which were primarily minimum pricing schemes, are of little concern here. They had, very likely, only a minor effect on the general profitability of coal mining. Violations of the price floors were widespread during the era of the NRA. The Guffey Act of 1935 was barely functioning when it was struck down by the Supreme Court. The first complete set of minimum prices promulgated by the National Bituminous Coal Commission, created under the Guffey Act of 1937, was hardly published before the defense program made minimum pricing a dead issue. After 1940 the war and postwar boom in business activity made outright governmental aid to the industry unnecessary.

It cannot be demonstrated statistically to what extent federal intervention promoted the recovery of the bituminous industry. While the New Deal assisted recovery generally, one of its significant contributions to the short-run stability of the industry was its sponsorship of unionism, for the union with its newly won power reduced the possibility of competitive wage-rate reductions among firms and regions. Future declines in coal demand could no longer be attended by "cutthroat competition" based on competitive wage rate reductions, because the union early succeeded in organizing nearly all firms. Moreover, the UMW used its power to eliminate by 1941 the principal wage rate differentials which had developed between the southern and the northern regions. This move in itself had an important effect on market relationships in the industry.[11]

The real matter for dispute at the end of the period was whether the intervention of the Federal Government had contributed to the

10. *New York Times,* July 7, 1935, p. 1.
11. See Chap. 7.

long-run stability of the industry. Few would disagree that the sponsorship of strong unionism by the government had helped to stabilize competitive relationships within the industry. But by 1950 many were wondering whether the solutions adopted to deal with the problems of the present would not create even greater problems in the future.

The years 1933–50 marked the resurgence of the bituminous coal industry. A variety of factors contributed to its recovery, particularly the outbreak of World War II. True enough, prices began to edge up as early as 1933. Employment and the number of active mines responded similarly. Firms continued to suffer losses, however, until 1940, though the amount of the net loss for the industry as a whole gradually declined as time passed.

A major contributing factor in the continuing financial losses was a rise in producing costs, especially labor costs. Wage rates increased rather substantially after 1933. More important, they rose more rapidly than did productivity. Since demand for the product was rising only moderately between 1933 and 1940, it is understandable that profits did not accrue to the industry as a whole until the defense program was well under way.

The war brought with it a sharp increase in coal demand. Prices and profits both rose. New mines were attracted into the industry in large numbers. Only the number of men employed declined, as a result of selective service demands and expanded job opportunities in other industries. But withal, bituminous output rose steadily.

Prosperity continued into the early postwar years as demand remained high—excluding 1948–49 when there was a decline in general business activity. Even so, by 1950 there was no evidence that the secular decline in coal demand had been or would soon be reversed. This did not bode well for the future of the bituminous industry.

CHAPTER 4

Policy Objectives of the Miners' Union

INTRODUCTION

THE WAGE POLICY of the United Mine Workers may be characterized broadly as "more and more—now." While this is a fair description of the union's long-run objective, it omits the subtler ramifications of union strategy. In reality, the UMW has directed its efforts over the years toward the attainment of three goals:

1. Steady improvement of the economic status of bituminous miners, particularly in relation to workers in other industries.
2. The stabilization of wage rates during downward cyclical fluctuations.
3. The stabilization of competitive relationships among firms and producing areas in the bituminous coal industry.

UNION POLICY FROM 1890 TO 1933

THE UNION BETWEEN 1890 AND 1919

The United Mine Workers of America was organized in 1890. Its antecedents have, however, been traced as far back as 1849. Since those earliest organizations of mineworkers were local in scope, they may be ignored safely in this brief review. The first national union of coal miners was organized in 1861, the bulk of its membership being concentrated in the mines of Illinois and Missouri. This union, dubbed the American Miners' Association, endured only briefly. In 1868 an unsuccessful strike severely weakened the group and brought about its early dissolution.[1]

Throughout the 1870's and 1880's several attempts were made to unionize large segments of the industry. The names of the unions were far more impressive than their achievements—the Miners' and Laborers' Benevolent Association, the Miners' National Association, the Amalgamated Association of Miners of the United States, the National Federation of Miners and Mine Laborers, the National

1. C. B. Fowler, *Collective Bargaining in the Bituminous Coal Industry* (New York, Prentice-Hall, 1927), pp. 36–7.

Trades Assembly No. 135, and the National Progressive Union.[2] For a variety of reasons—ineffectual strikes, factional disputes, inter-union rivalries—each of the organizations lasted but a few years.

The United Mine Workers of America was born on January 25, 1890. Its parents were the National Trades Assembly No. 135 of the Knights of Labor and the National Progressive Union, both of which had been contending for supremacy among the miners. Rivalry had cost the miners dearly, for the competing unions had been compelled to bargain on a statewide rather than on a nationwide level. The necessity of national wage bargains, to assure that wage rates in all unionized fields were related to one another, forced the competing unions into each other's arms.

The UMW made an auspicious start. Within a year of its organization 53,000 new members were added to the original 17,000. But the spurt in strength was only temporary. From 1893 to 1896, years of a sharp economic depression, the union apparently ran into membership difficulties.[3] By 1896 the UMW had only 10,000 members. Bold measures were required to save it. And such measures were forthcoming when Michael Ratchford called a national strike of miners for July 4, 1897. Lasting for 12 weeks and surviving a barrage of injunctions, the strike was won—mainly because the revival of business in 1897 brought an upturn in demand for coal. The union was saved and within a year membership had climbed to 33,000.[4]

The primary goal of the United Mine Workers in its early years, then, was its own survival. The desire to achieve recognition from the operators was of equal importance. Until 1898 the other objectives of the union were of little consequence compared to the need for organization and recognition.

The turning point in the union's affairs came in 1898, when a Joint Conference of miners and operators was held. It was at that conference and at another in 1902 that collective bargaining became an accepted principle in the coal mining regions of Illinois, Indiana, Ohio, and western Pennsylvania. The records of the conferences reveal very clearly the general lines of union policy.

The central theme of the union was the "principle of competitive equality." By this was meant that a system of fixed differentials in wage rates should be established. Wages should be "taken out of competition." It should be noted carefully that the union did not argue for uniform piece rates, but rather for uniform earnings. The two are not, of course, synonomous. In mines where the coal seam is narrow,

2. Ibid., pp. 38–42.
3. See Table 4-1 in appendix.
4. M. Coleman, *Men and Coal* (New York, Farrar & Rinehart, 1943), pp. 57–8.

low and/or laden with impurities, output per man will be comparatively low. Uniform tonnage rates in rich and poor veins would obviously penalize the worker in the poorer mine. The union insisted, therefore, that a thin-vein differential be established, with rates in thin-vein mines higher than in thick veins. Lower output would thereby be offset by higher earnings per ton, and in this way earnings of miners would tend toward equality despite disparity in the fertility of coal seams. Illustrative of thin-vein differentials were the rates in the Pittsburg area, where thin-vein rates were established at 8.53¢ per ton above the thick-vein rate.[5]

The mineworkers demanded also a machine differential. It was argued that output per man would be considerably greater for miners employed in mines using cutting machines. If rates in such mines were the same as in those using hand methods, machine miners would enjoy considerably higher earnings than hand miners. The union, therefore, insisted that the pick miner be protected by being paid higher per ton rates than the machine miner.

The amount of the machine differential was difficult to establish. The UMW had since its inception accepted fully the principle of mechanization.[6] Said union president John Mitchell in 1903:

> The unionists believe that machinery should be introduced with the least possible friction and the least possible hardships to individuals. When the employer is asked to increase wages or reduce hours, he frequently asks for an interval of a certain time in order to allow him to accommodate himself to the change, and the labor unions are now beginning to recognize the necessity of making great changes in industrial conditions by slow degrees. An equal duty should rest upon the employer to make alterations gradually, so as to extend the effect of the change over a series of years, and thus permit the workmen to accommodate themselves to the new conditions.[7]

This broad policy had at least two important aspects. First, there was a conscious effort to minimize the displacement of labor. In individual instances, for example, the UMW was able to convince mine owners that introduction of machinery should be accomplished over a period of years and adjusted to the normal rate of labor turnover.[8] Second, the union adopted in 1901 a wage policy that effectively

5. I. Lubin, *Miners' Wages and the Cost of Coal* (New York, McGraw-Hill, 1924), p. 89. As noted in Chap. 3, roughly 60% of the miners were paid on a piece-rate basis at this time. As machinery, particularly loading equipment, came into wider use, more and more mineworkers were paid by the day. At present, very few miners remain on piece rates. See below, pp. 66–7.

6. Ibid., p. 104.

7. J. Mitchell, *Organized Labor* (Philadelphia, American Book & Bible House, 1903), pp. 249–50, cited in WPA National Research Project, *Trade Union Policy and Technological Change*, Report L-8 (1940), p. 29.

8. WPA National Research Project, *Trade Union Policy*, p. 30.

slowed the rate of introduction of machinery: the demand that the machine differential be carefully regulated.

It was well known to both operators and miners that the introduction of cutting machinery would yield significant cost savings to the operators. Such savings would encourage further mechanization and would enable operators of machine mines to undersell their non-mechanized competitors. The pick mines would soon be placed in severe jeopardy. To protect the workers in the hand mines the union determined that a portion of the savings from machine mining would be diverted to the miners. In other words, the union agreed to a scale of machine rates which would be lower than pick rates, but not so low as to encourage precipitous installation of machinery. As John Mitchell put it in 1901: "Mr. Robbins [a coal operator] admits that the cost of machine mining as compared with pick mining is favorable to the operators. We contend . . . that there should be no advantage to the operator who owns a machine mine over the operator who owns a pick mine." [9]

The coal operators insisted that they should benefit from the lower costs arising from the use of machines. The union leaders argued, however, that the miners should at least share in the gains. They proposed that the machine rate should be equal to the pick-mining rate minus the cost of maintenance of the machine and a fair profit to the operators. The operators countered that the machine rate should be an amount which would yield an income to the machine miner equal to that of the miner working under a pick rate.

A compromise solution was devised. Machine differentials were established but they varied from district to district; in fact, the district conferences were delegated authority to work out machine rates for their respective districts. Thus, the machine rates for District 2 (Indiana) were established in the District Agreement of 1898 at 49.5¢ per ton or 75% of the pick-mining rate of 66¢. This agreement allowed Indiana operators a differential of 16.5¢ per ton as opposed to 10¢ per ton at Danville (Illinois).[10]

It will be noted that the union's objective in demanding thin-vein machine differentials was to equate or keep within bounds the differentials in earnings of all miners, without regard to where they worked.

Miners insist that they should not be compelled to suffer all the losses due to natural disadvantages. . . . They demand a high pick rate for thin-seam mines and they contend that even with this higher rate they finally bear part of the burden. Their earning capacity in the thin veins is necessarily

9. Lubin, *Miners' Wages*, p. 103.
10. Ibid., p. 106.

smaller than that of their more favorably located fellow workers, and a higher rate is necessary to enable them to maintain their standard of living.[11]

The differential policy is highly logical from the union viewpoint. Other things being equal, it lends itself admirably to single-minded pursuit by union officials. Given, that is, a strong union bargaining position and a generally profitable group of firms in the industry, it is the kind of policy that can be attained and enforced over a long period of time. Unfortunately for the union, neither of the "givens" applied during the early 1900's. There were important coal-producing regions in the Appalachian area which were not unionized and whose wage scales were often at marked variance with union scales in the Central Competitive Field. Moreover, within the unionized fields there were marked differences in total producing, administrative, and selling costs among the competing firms.

It was the differences in costs and the unstable pricing conditions that resulted therefrom that brought the operators together in joint conference. A coal operator attending the Illinois Joint Conference of 1902 declared: "We have banded together here so that the operators in every district might exist, notwithstanding the different conditions that prevail. . . ."

The operators demanded that wage rates be fixed for all firms in the Joint Interstate Conferences on a basis that would allow every operator to do business. Each operator, it was argued, must be given a chance to mine his coal and get it to his market in competition with all other operators. "So long as we work on those lines some miners will have to accept less wages than others." [12]

The miners accepted certain of the operators' proposals. A major concession was the union's acknowledgment of the effect of existing freight-rate differentials on the coal wage structure. As has been noted, freight charges represent a considerable proportion of wholesale selling price. In recent years, for example, average freight charges have equaled or exceeded the total producing, administrative, and selling costs of the coal operator. Where freight rates vary among operators selling in the same market, it is obvious that sellers paying higher rates may be placed at a distinct competitive disadvantage. That is particularly true of mines whose producing costs are high compared to those of their competitors.

The operators argued that the wage structure should reflect the existing scale of freight rates. This was agreed to by the UMW. Illustrative of the arrangement was the schedule of pick-mining rates

11. Ibid., p. 95.
12. Ibid., p. 72.

in central and southern Illinois. The pick-mining rate for the worker in southern Illinois was 4¢ less than that in central Illinois, "a condition that is said to owe its origin to the fact that the cost of transporting coal from the former field to the principal market [Chicago] is greater than for the central field." [13]

It will be recognized that the assumptions underlying the freight differential are diametrically opposed to the argument that the mining rate must allow for variations in mining conditions through a thin-vein differential. The latter differential increases the cost disadvantage of the less favorably endowed mines. The freight differential, on the other hand, works to reduce cost differences among mines, and works against uniformity.

Why did the union agree to the system of freight differentials? It could be argued that the union was concerned lest certain mines be forced out of the industry with an obvious impact on employment of union members. Rising unemployment among the membership might well have endangered the stability of the union, just recovering from the severe impact of the depression years of the 1890's. Or the union may have feared that too ambitious a position on wage uniformity might have caused a breakup in the newly won collective bargaining arrangement which embraced several states.

It is known that the union resisted certain differentials altogether and objected to others on the grounds that they were not sufficient. The latter were agreed to only because the union's bargaining position was not sufficiently strong. That is revealed in the 1920 testimony of Mr. John L. Lewis, appearing before the United States Coal Commission:

> Inequalities in tonnage rates . . . prevail. Some of these conditions have prevailed for a great number of years. They have prevailed because the mine workers have heretofore been powerless to secure their modification or adjustment and though in conference after conference we have taken up these matters and sought consideration, we were oftentimes compelled to enter into an agreement which did not carry with it any adjustment of these differentials and startling inequalities. . . .[14]

The efforts of the UMW between 1890 and 1920 were not limited to gaining equality in wages. The union was also interested in changing certain practices which reduced workers' earnings. Among these practices was the varying size of work rooms. Where the rooms were wide (more than 20 feet in width) the distance from the sidewalls to the mine car made it impossible to shovel new-cut coal directly into the cars. It was necessary to pile the coal midway between the wall

13. Ibid., p. 89.
14. Berquist and Associates, *Economic Survey*, p. 357.

and the track, then shovel it into the car. The union demanded that a standard room be defined and that additional compensation be fixed for rooms over or under the standard width.[15]

Added compensation was sought for miners working in mines with sloping floors. Extra work was involved in pushing loaded cars into the main haulageway. Such work reduced the time available to the miner for the direct production of coal. The union insisted, too, on special payments for cutting new passages and air courses. Live entries and air courses are comparatively narrow; the labor involved yields less output than the same effort would yield in a room. The difference in output is so marked that a supplementary payment called "yardage" is made for each yard of distance driven.[16]

One of the long-standing sore points between the union and the operators was the union's insistence that miners be paid for all coal brought to the surface. For many years payment had been made only for marketable coal, meaning that coal which had first been screened. Many coal users at that time were interested only in lump coal, which meant that the finer sizes had to be separated on the screens and disposed of at virtually distress prices. Thousands of tons of coal went through the screens, tonnage for which the miner received no payment.

The union insisted that all of the miner's output be paid for, that tonnage rates be paid per ton "run of the mine." The operators steadfastly maintained that this procedure would be impractical in that it would work severe hardship on the small mines. They were concerned particularly for the fate of the nonmechanized mines, whose total output was small relative to their larger competitors and whose marketable output was even less. To force the small firms to pay on a run-of-mine basis would jeopardize seriously their continued existence.

In 1912 the union played a large role in the passage of a run-of-mine law in Ohio. But adoption of the run-of-mine system was delayed in part by disagreement within the union itself on the methods for determining the basis of payment, and there was continual disagreement between miners and operators on the kinds and sizes of screens to be used. Not until 1916 did the union achieve full adoption of the mine-run basis of payment.

Through most of the period 1890–1919 the union was a growing organization. During the general depression of 1893–96 wage rates were first stabilized, and later, under the stimulus of the war, rose rapidly. Membership increased from a low of 33,000 in 1897 to more

15. Lubin, *Miners' Wages,* p. 118.
16. Ibid., p. 119.

than 400,000 in 1917.[17] Furthermore, the benefits obtained for union members could readily be observed. As was pointed out in Chapter 3, wage rates paid to union miners were somewhat higher than those paid to nonunion miners. Even after the rapid increase in both union and nonunion wage rates during the war, union rates were still higher than in the unorganized areas.

Yet there were fundamental weaknesses in the union position. The nonunion areas were steadily cutting into the markets of the organized mines. Nonunion mines were working more steadily as a result of the shift in tonnage from the unionized north to the nonunion south. The war temporarily obscured this and other problems. But after the war they reappeared, marking the beginning of a disastrous period in the history of the UMW.

THE UNION BETWEEN 1920 AND 1929

The years 1920 to 1929 were marked by a sharp reduction in union strength. There were losses in membership. Wage rates in the unionized fields declined. Many firms with union contracts reverted to the "open shop."

In the first years after World War I there was labor strife in the industry. The union was eager to throw off the restrictions on wage demands which had been necessitated by mobilization. With prices and profits soaring at the war's end the union evidently reasoned that the time was appropriate for substantial wage increases. Accordingly the union in 1919 demanded, among other things, a 60% wage increase, a six-hour day from bank to bank,[18] a five-day week, time-and-one-half for overtime, and double-time for Sundays and legal holidays.[19]

After a major strike and the subsequent intervention by the President of the United States, a compromise was arranged under which all classes of mine labor received a wage increase averaging 27%. The other major demands were not met. Moreover, the agreement was not to expire until March 31, 1922.

The war-born prosperity of 1918 to 1920 faded rapidly after late 1920. Throughout 1921 output of coal declined and employment fell with it. In particular, the shift in production from northern to southern fields—remarked in Chapter 2—resumed its prewar trend, bring-

17. Coleman, *Men and Coal*, p. 92.

18. This includes travel time from portal to portal.

19. Fowler, *Collective Bargaining*, p. 84. In 1919 Mr. John L. Lewis became the acting president of the UMW. It is quite possible that the demands of the union were motivated by the new president's desire to prove his mettle as well as by the desire of the union to maximize the workers' earnings.

ing in its wake relatively heavy unemployment in the unionized fields of the north. The blame for the northern losses was laid at the door of the union, chiefly on the grounds that high and inflexible wage rates were hurting northern operators relative to their southern, nonunion competitors.

There began an apparent effort to break up the collective bargaining machinery that had grown up in the industry, particularly in the Central Competitive Field. Many operators proposed to make substantial (up to nearly 50%) wage reductions. Many refused to continue the "check-off." Their position was revealed in statements such as this one:

The fixed belief, held quite generally by coal men throughout the country, [is] that mine labor as well as organized labor in other industries are demanding negotiation in unwieldy units. Adequate and proper representation for wage agreement negotiations for such large areas give a conference body of such size and diversified opinion that the original intent and purpose of collective bargaining is made impossible.

This is not a new idea, either, with the operators. It is based on ample precedents of which the public has apparently already approved. The United States Steel Corporation, through Judge Gary, refused to negotiate with others than his own employees. . . . Nor could there be given any clearer evidence than the recent demonstration of railroad labor of the futility of such massive and country-wide effort on the part of organized labor attempting to force the continuance of an unnatural wage level through sheer strength and an alleged ability to paralyze all industry through the use of so-called "economic power." [20]

The miners' reply was a vow that no separate agreements would be made until the basic agreement for the Central Competitive Field had been settled. When the issue became stalemated, the union struck in mid-1922. Violence ensued in certain areas. But revived demand for coal soon caused a large number of operators to agree to an extension of the contract to March 31, 1923. This proved to be an opening wedge in forcing the operators of the central and outlying fields to sign new agreements on the interstate basis.[21] Later an agreement was reached under which the 1920 rates were re-established and the interstate agreement retained for a period of two years.

This contract set the stage for the famed Jacksonville Agreement of 1924. Under this contract the operators in the Central Competitive Field agreed to continue to pay $7.50 per day. This wage scale was considerably above the rates prevailing in the nonunion areas of the

20. *Coal Age,* March 2, 1922, p. 381.
21. Fowler, *Collective Bargaining,* p. 101.

south. What is more, the shift of production from north to south was continuing at a steady pace. Unemployment was rising in the northern mines.

Since the union was unable to organize mines in the south, the UMW's alternative seemed to be to accept a lower wage scale. But the announced policy of the United Mine Workers was "no backward step." This was no idly made observation. The evidence strongly indicates that the union president, John L. Lewis, had concluded that market conditions would force many operators to close down their high-cost mines. While this would increase the number of unemployed union miners, it was felt that greater stability would be realized in the industry if there were fewer mines and miners. Furthermore, Lewis contended that the higher freight rates for the long hauls from the nonunion territory would offset the higher wages paid in the north.[22]

The position of the union was summarized well by Lewis himself in 1925: "The bituminous industry is suffering . . . the pains incidental to a long-delayed adjustment. . . . When it is complete, there will be fewer mines and miners and it will be a prosperous industry." [23]

The Secretary of Commerce, Herbert Hoover, with whom Lewis had consulted on the whole problem, was even more sure of the benefits which would flow from the Jacksonville Agreement. In his annual report for 1924 he declared:

> The coal industry is now on the road to stabilization. The benefits lie not only in the provision of coal to the consumer at lower prices than have been attained at any time since the beginning of the war. The gradual elimination of high-cost, fly-by-night mines is bringing about a greater degree of concentration of labor upon a smaller number of mines, the increase in days of employment per annum, and thus a larger annual return to the workers. The inherent risks in the industry will be decreased because the efficient and stable operator will no longer be subjected to the type of competition that comes from those mines that exist only to take advantage of profiteering periods.[24]

The predictions were none too accurate. The shift of output toward the southern fields continued. In 1920 about 29% of the nation's coal had been mined in union fields and 25% in nonunion, with the remainder unaccounted for. In 1924 24.6% was union coal and 36% was nonunion. Machine-cut tonnage was 7% greater in 1924 than in

22. Coleman, *Men and Coal,* p. 124.

23. C. D. Edwards, "Coal Unionism," in *American Labor Dynamics,* ed. J. B. S. Hardman (New York, Harcourt, Brace, 1928), p. 182.

24. *Twelfth Annual Report* (Washington, Government Printing Office, 1924), pp. 13–14.

1940. Mines in the union fields were shutting down but some were reopening on a nonunion basis. Other mines were openly violating the Jacksonville Agreement.[25]

As a complement to its policy of "no backward step," the union renewed its efforts to organize the southern coal fields. Injunctions, yellow-dog contracts, blacklists, and armed guards smashed the attempt. By 1926 the Jacksonville scale existed in name only. The union scale committee was authorizing the miners to make district agreements as best they could. Only in Illinois was the union scale of $7.50 still in effect. And in 1927, when the Jacksonville Agreement formally expired, the UMW reluctantly authorized its Illinois locals to accept day rates as low as $5.00.

The dues-paying membership of the UMW steadily melted away during these years. Figures which were questioned as to accuracy but not refuted by the UMW indicated that by 1929 only 84,395 members of the UMW were paying dues, a far cry from the 400,000 dues-paying members of 1920.[26] The conservative Mr. Lewis turned to government for assistance. By 1932 he was ready for governmental regulation of the coal industry, and with the counsel of the UMW he prepared to write a draft bill.[27]

Throughout the period 1920–29 the UMW was the prisoner of events, rather than the shaper of them. Intense interregional and interfirm competition in the 1920's reduced selling prices. With labor cost so large a percentage of total cost, it was inevitable that falling prices should exert downward pressures on the wage level. Had the industry been fully organized, the union might have hindered the decline in wage rates, though very likely at the cost of some additional unemployment. The industry, however, was only partially unionized. Moreover, the nonunion areas were able to sell a better quality of coal. The two factors, flexible wage rates and better coal, gave the nonunion areas an overwhelming advantage in underselling their unionized competitors. Willingly or not, the unionized operators were compelled to ignore the more rigid union scale.

The UMW's plan to drive mines out of the industry by maintaining a level of wage rates higher than in the nonunion areas thus failed. Many operators simply closed down their mines and reopened in nonunion territory. High wage rates thus complicated the union's effort to stabilize the industry. But the union could not have agreed in advance to lower wage scales in the unionized fields, since this would have been a concession of defeat without even giving battle. That

25. Coleman, *Men and Coal,* p. 125.
26. Ibid., p. 139.
27. Ibid., p. 146.

would have been, in turn, a clear admission by the union leaders of their own ineptness, an act of political suicide for the union president. The union had, as a result, no choice but to drift helplessly.

THE UNION BETWEEN 1929 AND 1932

Badly weakened by the persisting interregional competition of the 1920's, the UMW was powerless to combat in any way the impact of the general cyclical downturn which began in 1929. Employment, prices, and profits continued to fall. The principal difference—and an important one—between the 1920's and the early 1930's, in fact, was that in the later period the demand for bituminous coal diminished in absolute terms, where previously it had remained at substantially the same level as in 1920.

The UMW's only recourse during the early depression years was to seek the aid of the Federal Government. Throughout those years, as will be recounted later, the UMW pleaded for some form of public regulation of the coal industry. In this it was unsuccessful, primarily because a cardinal union demand was that any governmental intervention must carry with it a guarantee of the principle of collective bargaining. To this the operators were adamantly opposed.

Union Policy since 1933

IMPROVEMENT OF THE MINERS' STATUS

Coal mining has never been an attractive occupation. The work is physically demanding, hazardous to life and limb, and, until recently, ill paid. Moreover, coal miners had virtually no social status. In brief, they were economically, geographically, and socially isolated. To them "the union is more than a collective bargaining association, it is the pillar of their hopes. As long as the union is preserved they are not serfs, they retain a glimpse of freedom and an awareness of potential power. The fortunes of the union are completely entwined with their own personal histories." [28]

The broad objective of union policy has been to raise the economic and social status of the miners. Internal political considerations play a part in this orientation of policy. But political factors seem to be of secondary importance now that the leadership has become so entrenched. The particular goals now pursued by union officials seem to be based on economic considerations, founded as the policies are on an unwavering (though incorrect) analysis of the causes for the recurring crises in bituminous coal mining.

28. J. A. Wechsler, *Labor Baron. A Portrait of John L. Lewis* (New York, Morrow, 1944), pp. 8–9.

The wage strategy of the United Mine Workers has three aspects. First, the union seeks to raise the wage incomes of its members. Second, it attempts to stabilize occupational wage-rate differentials. Third, it desires to equalize wage rates between competing coal regions.

Secular Increase in Wage Incomes. No effort need be spent debating the arid question whether the UMW is interested in a secular increase in wage rates. Common sense, amply supported by history, makes that obvious.[29] What is more important is that the union has sought to raise over time the wage incomes and working conditions of its members. This has been accomplished in several ways. First, the level of wage rates has been increased steadily since 1933. Second, the union has demanded reduction in the hours of work as productivity has increased. Third, it has insisted on constant improvement of safety conditions on the job. Fourth, wage increases have included a variety of "fringe" benefits: paid vacations, a noncontributory welfare fund, portal-to-portal pay, and so on.

All these objectives add up to a steady improvement in the status —both economic and social—of the coal miner. There can be no doubt that such is the primary motive of the union. The president of the UMW rejects the suggestion that wage gains of his union will be wiped out by rising prices. His clear position is that the union can and will assure that the economic status of UMW members is constantly improving in comparison with other workers, both union and nonunion.

During an interview Mr. Lewis sardonically told me that he hoped I was not "one of those economists who believe that wages and the cost of living rise together." He went on to say that the UMW had recently analyzed conditions in "one of our great coal mining states"; that the study showed that over a 50-year period coal prices (f.o.b. mine) had risen by 40% while wages had increased 1,700%. With obvious pride he pointed out, too, that once a miner lived in "a shack with a dirt floor. Now he has a fine home with a floor and a rug; a car or even two, television—and his wife has taken her rightful place in society. *That* is what the coal miner wants."

Stabilization of Skill Differentials. The United Mine Workers has, like many another union, attempted to assure for its members reasonably equal incomes. This does not mean that the union demands equal wage rates for all skills of mine labor. On the contrary, the UMW countenances—undoubtedly with the support of its membership—a complex structure of skill differentials. There is one set of rates for inside day workers, another, somewhat lower, for outside

29. Changes in wage rates since 1933 are discussed in Chap. 7.

employees. Within each classification there are variations based on the specific occupation. Moreover, the union only recently won from the operators its long-standing demand for shift differentials—extra payments for men on the second and third daily shifts.

There is little indication that the UMW entertains any serious wish to alter importantly the range of occupational differentials. Since 1933, however, two changes have been effected in the national agreements. In 1943 the union sought and received an increase of 85¢ per day for the lowest-paid inside day workers, the increase being over and above the amount asked for all employees. In 1945 the earnings of inside workers were increased by $1.50 per day while those of outside workers rose only $1.07 per day. The latter change increased the spread between the highest- and lowest-paid men to $3.75 per day.[30]

It is improbable that these two adjustments indicate any fundamental interest on the union's part in altering the structure of occupational rate differentials. There is no evidence that the union's leaders have taken any further steps along these lines in recent years, probably because there have been no complaints of any consequence from the membership. It must be presumed, therefore, that adjustments will be sought as and when deemed advisable, in terms of other objectives of the union.[31]

Equalization of Regional Differentials. The UMW, despite some confusion in the records, long has sought to make uniform the structure of wage rates in the several coal regions. It is now clear that the differentials acceded to by the union in the wage arrangements made in the early 1900's were the result of competitive conditions in the industry, together with the inferior bargaining position of the miners' union. The early bargains were, in other words, the best the union could obtain.

At bottom most wage differentials have two chief reasons for being. They result from severe economic pressures or they are historic survivals from unregulated, non-union conditions. Given coal seams in backward districts with a population living on low economic levels, the tendency is for coal mines . . . to build up a wage system adapted to the current standard of living. Thereafter, the decrease in production costs resulting from full and continuous production induces the operator to push his sales and shipments to the most remote markets he can reach. This results in establishing an equilibrium between costs and realization at his low wage standards. When the union enters this situation, it can only gain recognition by accepting the current wages.[32]

30. See Hearings, Senate Committee on Banking and Currency, *Economic Power of Labor Organizations,* 81st Congress, 1st Session (1949), Pt. 1, pp. 310–11.

31. It seems likely, for instance, that the 1943 demand represented an effort by the leadership to "get something" from the War Labor Board, which only recently had limited wage rate rises to the "Little Steel formula."

32. Berquist and Associates, *Economic Survey,* p. 406.

Throughout the 1920's, as before World War I, the union was powerless to eliminate the differentials. Meantime, the competitive wage cutting that accompanied and abetted interregional competition exaggerated the range of differences in wage rates. Moreover, the union was so occupied in its effort to maintain the general level of wages that it could not devote much effort toward elimination of regional differentials.

Its great opportunity to do so came in 1933. New Deal legislation gave the union the chance to build its bargaining power; the pro-union attitudes of some NRA officials gave quasi-legal sanction to the union's efforts. The first Appalachian wage bargain, effective October 2, 1933, provided for daily wage rates which in northern West Virginia were 24¢ less than in Ohio and Pennsylvania; in the other southern states the daily wage rate was 40¢ less than in the north. This evidently was the best bargain the union could obtain at that time. But in April 1934 the union succeeded in wiping out the 24¢ differential in northern West Virginia. The north-south differential of 40¢ per day succumbed in 1941 after prolonged and bitter attack.

This much of the history of the UMW's struggle for "equal pay for equal work" seems straightforward. The apparent ambiguities crop up when the discussion is turned toward tonnage or piece rates, as distinguished from daily rates. The written records abound with union testimony that piece-rate differentials are inevitable and necessary. Even at present the basic Appalachian contract (originally concluded in 1941) includes an elaborate structure of tonnage rates with marked differentials in favor of southern mines. Moreover, the differentials have been left almost unchanged since 1933. Increases granted to pieceworkers since that year have been of equal amounts, so that the absolute differentials are unaffected.[33]

Philip Murray, then vice-president of the UMW, stated the union's position on piece-rate differentials in 1935:

In the final determination of what proper wage relationships should be in these various districts, consideration is given to every element of cost. . . . Charts, maps, production figures, earnings, costs and every single factor having to do with the operation of a coal mine is given consideration in the final determination of what differential conditions should prevail between competing districts. . . . The yardstick of common sense is applied so that no interest will be injured, no dislocations of tonnages will take place, no disturbances of production will take place, as a result of the wage arrived at in conferences of this description.[34]

33. The rate structure and changes in it between 1935 and 1941 are recorded in Table 4-2 in appendix.

34. Testimony in *Carter v. Carter Coal Co.,* Verbatim Transcript of Proceedings before Supreme Court of District of Columbia, pp. 525–6.

In like fashion the union president laid down the broad policy of the union at the Appalachian Joint Wage Conference of 1934:

On the question of differentials . . . the mine workers . . . recognize . . . that there had grown up in the industry, inequitably if you please, certain conditions for which neither the operators nor the miners are responsible. They recognize the commercial problems and they recognize that in dealing with these questions the millennium cannot be reached at once, and it takes time and study and careful consideration of these problems to ascertain and develop the correct relationship between the various fields. The mine workers can only say that . . . they approach this question from the standpoint of rendering equal justice to the operators of the several fields and the mine workers employed in those areas.[35]

The ambiguity of union policy in the matter of wage-rate differentials is not so real as it appears. In the light of hindsight it seems certain that the union was merely marking time. Payments by the ton are now virtually an anachronism in the coal industry. The introduction of loading machines which vastly increase productivity have made it impossible to continue a piece-rate system. For the coal operators time payments (by the day, week, or month) make labor cost a more predictable quantity. Moreover, the incentives allegedly provided by piece rates become inconsequential when output becomes dependent primarily on the physical capacity of a machine. Finally, the installation of machines permits the use of crews of men who move their equipment from room to room. Supervision becomes feasible, which it was not when individual miners worked alone, each in his own room. With supervision, incentive payments become even less meaningful.

The union sees the problem in a different light. "Labor productivity with power-driven machines is a variable of the design, of the management, of the planning and so on. Therefore, to set a tonnage rate for mechanized mining is a task that can hardly satisfy either labor or stockholders. If management is lax, conditions will reduce a man's output and he will suffer through no fault of his own (lack of supplies, interrupted power, unprepared working place, etc.)." [36]

As more and more mines have become mechanized, fewer and fewer miners are being paid on a piece-rate basis. The tonnage-rate differentials have declined in significance in direct proportion to the decrease in hand loading. The differentials remain in the contract pretty much as a relic of a bygone day. Meantime, most of the remaining hand-loading mines are becoming mechanized and are adopting time rates

35. UMW *Journal*, 45 (March 15, 1934), 5.

36. S. H. Slichter, *Union Policies and Industrial Management* (Washington, Brookings, 1941), p. 297n., quoting a personal letter from a union official.

—which are uniform in all regions. If the union did not plan to finesse piece-rate differentials in this way, it did at least try to hurry the process along. For the UMW has, at least since 1933, been an ardent advocate of mechanization.[37]

Why did not the union simply demand the elimination of *all* differentials instead of biding its time while mechanization took place? In fact, the union repeatedly made this demand. Each year until very recently its initial proposals to the operators included a provision that "existing inequitable differentials within and between districts be eliminated." [38] It is doubtful, however, that much importance should be attached to this. The published records of the joint wage conferences reveal no debate on the subject.[39] This suggests that the union raised the issue as a bargaining point, as something to be yielded in return for something else wanted more urgently.

More important, the union leadership reasoned that time—and the pressure of rising wage rates on producing costs—was on the side of increased mechanization in the industry. What other reasonable interpretation can there be of Lewis' remark that "the mine workers . . . recognize that the millennium cannot be reached at once"? What point in precipitating a pitched battle over an issue that was less compelling than the then urgent matters of the elimination of the north-south differential and the extension of the union shop to all mines in the industry?

The conclusion seems justified that the union's intent is to equalize wage rates for similar skills, regardless of the location of a mine or the physical conditions of mining. Moreover, the union's goal is close to complete realization. As will shortly be seen, however, this policy has ramifications far beyond the simple achievement of income equality among mine workers.

PROTECTION OF WAGE RATES AGAINST CYCLICAL FLUCTUATIONS

The UMW has an abiding horror of cyclical downturns.[40] Declines in product demand and the (derived) demand for labor have resulted in the past in a general reduction of wage rates in the industry. The

37. See below, pp. 71–2.

38. See, for instance, wage proposals adopted on February 28, 1945, as printed in UMWA, *Proceedings of the 39th Convention* (1946), p. 59.

39. See, for example, the transcribed records of the 1945 and 1946 conferences, printed in *Proceedings of the 39th Convention.* On the other hand, the opinion of the fact-finding panel of the War Labor Board in the 1943 coal wage dispute declared that "existing differentials are based on costs that are so deep that it would be impractical to consider any change at this time." UMW *Journal, 54* (June 1, 1943), pp. 6–7.

40. The "depression consciousness" of the union is discussed in detail in Chap. 9.

union's power has suffered in direct proportion to the magnitude of the depressions. Moreover, depression-induced reductions in wage rates have cost the union much effort to recoup the losses when demand for coal later revived. The UMW's strategy, therefore, has been to insulate miners' wage rates against cyclical shocks, to make them rigid in the downward direction.

As might be expected, this policy has evoked sharp criticism from many quarters within and without the union. It is stoutly argued that wage rates can be maintained, given a decline in product demand, only at the cost of increased unemployment. The UMW counters on two levels. First, it denies that a lower level of wage rates would appreciably increase sales of the product. The experiences of the 1920's and 1930's have demonstrated to the union's satisfaction that wage-rate reductions lead only to further price and wage cuts. Neither the miner nor the operator benefits from this process. The UMW maintains, in the second place, that the price of coal is a very small part of producing costs for the large coal consumers. Price shadings permitted by a decline in wage rates will not cause any significant change in consumption, i.e., the demand for coal is relatively inelastic.

The union's position, therefore, is that prosperity can be restored best by steady or preferably rising wage rates. Said John L. Lewis during the depression of the 1930's:

> . . . A wage increase immediately increases the consuming power of the people, and increased consuming power means lower production costs, lower prices and probably increased profits to the employer because of greater sales. . . . [Furthermore], a wage increase is a stimulus to the adoption of the best technical processes. The best argument for the capitalist system is that competition provides a needed stimulus to efficient work and efficient productive processes.[41]

The merits and demerits of this argument aside, it remains that a policy of rigid wage rates during cyclical declines makes the greatest political sense to the union. Its achievements on behalf of its constituents are most readily measured by the level of wage rates. On the other hand, as A. M. Ross has commented, "the real employment effect of the wage bargain is lost in a sea of external forces. The volume of employment associated with a given wage rate is unpredictable before the fact, and the effect of a given rate upon employment is undecipherable after the fact. The employment effect cannot be normally the subject of rational calculation and prediction at the time the

41. J. L. Lewis, "Effect of Moderate and Gradual Wage Increases on Prices and Living Costs," *The Annalist*, 50 (November 12, 1937), p. 779.

bargain is made, and union officials are normally in no position to assume responsibility for it." [42]

STABILIZATION OF RELATIONSHIPS BETWEEN REGIONS AND FIRMS

The troubles of the 1920's and 1930's left a deep impression on the union leadership. All the key figures in the UMW at present were, in fact, the policy makers during the earlier period. It is understandable, then, that they should try to remedy what to their minds were the causes of the crisis in coal during the twenties. Believing that the problems of the industry were caused not by fluctuations in demand but by "excess capacity," the leaders of the UMW decided to eliminate the "extra" mines. The decision was that high wage rates should be imposed on all firms—including those in the nonunion south. But not until the passage of the National Industrial Recovery Act was the union able to organize workers in the southern fields.[43]

The other string in the union's bow was the equalization of wage rates in all fields. No firm was to be permitted to remain in operation because of wage rates lower than its competitors. The union made no effort to conceal its plans. Said its president in 1933:

> There is no justification for the maintenance of a system which permits the present differentials, either in or between districts. . . . Lack of uniformity of wage and tonnage rates between competing districts has pauperized the miners and bankrupted the operators. . . . In the . . . [NRA] coal code . . . we have agreed to a 5 per cent differential for the southern fields. In doing so, however, we are merely making a concession to a traditional practice, and do not admit that such a differential is justified by differences in the cost of living.[44]

Lewis's implication was unmistakable that the UMW had agreed to the north-south differential only as a tactical maneuver, that the

42. A. M. Ross, *Trade Union Wage Policy* (Berkeley, University of California Press, 1948), p. 80. J. R. Hicks has pointed out, too, that "to the Trade Unionist wages and unemployment naturally appear to have little connection. The initial unemployment may be too small to be really noticeable; and the later additions are most easily ascribed to quite different causes. That which comes from substitution is put down to 'labor-saving machinery'; that which comes from bankruptcy is ascribed to the inefficiency of employers. That the wage policy which has been going on so long and has seemed so successful has anything to do with present calamities seems too far-fetched to be considered." J. R. Hicks, *The Theory of Wages* (Gloucester, Mass., Peter Smith, 1948), pp. 184–5.

43. So anxious was the union to organize the south that it emptied its treasury in 1933 to send out organizers even before the NIRA became law. Shortly after the bill was enacted, the UMW claimed 340,000 members in the Appalachian fields and 170,000 in West Virginia alone. *New York Times,* September 26, 1933, p. 7.

44. Statement at NRA Coal Code Hearing, *New York Times,* August 11, 1933, p. 4.

union would ultimately try to wipe it out. Moreover, there can be no doubt that uniform and rising wage rates were to be the instrument for eliminating the marginal firms. "The United Mine Workers . . . in supplying a uniform competitive labor cost for bituminous coal mining, has been the only stabilizing force which the industry has ever had." [45]

The elimination of excess capacity through union wage policy is based on uncomplicated reasoning. Labor cost in coal mining is a high percentage of total cost. Its amount differs from mine to mine, varying inversely with the volume of output per man. The less efficient firms are thus put at a cost disadvantage if their wage bills are equal to the wage bills of their more efficient competitors. If demand for the product is unchanged or falling, the high-cost firms will be squeezed between rising cost functions and falling margins. They may try by mechanization to offset their rising producing costs. But if other firms are also installing machinery, the relative cost position of the marginal firms will at best be unimproved. If the high-cost mines are unable because of geological conditions to introduce the most modern equipment, their competitive status will be worsened.

The union's strategy, then, is to eliminate all wage-rate differentials. The level of wage rates will at the same time be pushed up. The effect of the two policies will be the encouragement of mechanization, the cost benefits of which will be "shared" with the miners. At the same time, rising costs will slowly force the relatively inefficient mines out of the industry. This will mean that the remaining firms will absorb that part of the market which has been served by the deceased mines, enabling the survivors to operate on a more regular schedule. Product prices will be "firmed up" by the elimination of those mines which seek to survive by selling "below [total] cost."

When, therefore, the coal producers protest that the effect of the union's demands will be to drive many firms out of business, it is understandable why the union is unmoved. That is just what the union wants.[46]

THE LEVEL OF EMPLOYMENT

The union, it has been noted, has an abiding belief that excess capacity lies at the heart of the coal industry's problems. The union sees the issue as one of too many mines and too many miners. The remedy is largely, in the union view, the elimination of the "excess" of both.

45. J. L. Lewis, ibid., p. 24.

46. As will be pointed out in Chap. 5, that is also what the most efficient firms want. It explains in large part why the big coal companies have frequently led the way to general wage settlements.

The UMW approaches the problem of reducing employment in several ways. First, it has been an active proponent of child labor legislation. Its contract with the operators includes a provision that no boys under 17 or 18 years of age (depending on the relevant state law) will be permitted to work in the mines. Second, the union enforces strictly the seniority provisions in its contract, thus tending to reduce the rate of entry of younger men into the industry. Third, its encouragement of mechanization tends to reduce mine labor requirements. Though few men are actually thrown out of work by machines, there is little need in a mechanized mine to replace miners who are annually retired, disabled, or killed. The rate of displacement, in other words, is roughly balanced by the rate of human attrition. Finally, union policies which tend to force marginal firms out of the industry will tend also to reduce the volume of employment.[47]

There should be no assumption from these observations that the union attempts to calculate the "employment effects" of its wage demands. Several variables—notably product demand—affect the volume of employment at any given time, making predictions of the effects on employment of wage-rate changes virtually impossible. Yet union officials freely admit that they "planned long ago on a steady reduction in employment as time went on." [48] Their strategy is this: fewer miners using more machines will command ever higher wage rates because of increasing productivity. And with a declining number of mines in operation, the remaining firms—and their employees—will enjoy more regular working time and, perhaps fewer hours of work per day.

TECHNOLOGICAL CHANGE

John L. Lewis has been called, among other things, "the best salesman [the machinery industry] ever had." [49] The implication is, of course, that rising wage rates have caused coal producers to substitute machinery for labor.

There seems little room for doubt that the United Mine Workers wants the mines mechanized. The union president said in 1947:

> The UMWA . . . [takes] the position that the only way in which the standard of living could be increased . . . would be by increasing the productivity and lowering the unit costs and utilizing the genius of science

47. There can be no certainty of this, since men displaced from one mine may be employed elsewhere in the industry.

48. Union officials told me, too, that they took "everything into account"—including the employment effects—when formulating their wage demands. This seems doubtful for the reasons given.

49. "Continuous Coal Mining," *Fortune* (June 1950), p. 118.

and the automatic machine . . . and the usage of power to do the work of human hands, and the UMW educated its membership through the years to an acceptance of that policy. . . .

As the result of that policy, the UMW . . . declared that the miners had a right to participate, through increased wages and shorter hours and improved safety and better conditions, in the increased productive efficiency of the industry, holding that there were three parties to the profit by increased proficiency and greater production: (a) the investor who was given a larger return and a more secure investment; (b) the mine worker who would get higher wages and shorter hours, improved safety conditions; and (c) the consumer of the product who could buy this product at a lower unit cost. . . .[50]

The union will accept mechanization so long as it shares in the benefits accruing initially to the operators. This is a long-standing policy, first appearing when machine differentials were being established in the early 1900's.[51] There have been rumblings from the ranks against mechanization, particularly during the depression of the 1930's. The objections were easily overruled.[52]

The union's current strategy is easy to describe. By raising wage rates (and labor costs per ton) and by eliminating regional wage-rate differentials, heavy pressure to mechanize will be brought to bear on all firms, especially the relatively high-cost operations. Since the union insists that it must share in the benefits of mechanization, the increased use of machines will enable the union to exact even higher wage rates or shorter hours or both. Higher wage rates, in brief, encourage mechanization, which permits still higher wage rates. Seemingly the cycle could go on indefinitely, limited only by the financial capacity of firms to install machines and by the ingenuity of mining machinery manufacturers to devise ever more efficient equipment. The actual limits are set, of course, by the level and elasticity of demand for coal.

COMPETITION WITH SUBSTITUTE FUELS

Despite charges to the contrary, the union is not insensitive to the relative decline in consumption of bituminous. Nor does it minimize

50. Hearings, House Committee on Education and Labor, *Welfare of Miners,* 80th Congress, 1st Session, *1* (April 3, 1947), p. 41.

51. See above, pp. 53–5.

52. S. H. Slichter points out: "At the convention of the union in 1936 . . . there were [many] resolutions [which] . . . proposed that taxes be imposed on loaders . . . At the convention in 1938 many resolutions proposed taxes on labor-saving equipment, and at the convention of 1940 no less than 29 resolutions on machinery were introduced, most of them proposing a special tax on it. The officers of the union opposed these proposals and advocated accepting mechanization and attempting to extract benefits from it." (*Union Policies,* p. 272.)

the effects of falling product demand on the demand for mine labor. The UMW's chief difference with the coal operators in this matter is on the causes for the decline in demand and on the policies to be devised to combat it.

Union leaders are unmoved by the plea that bituminous coal is being "priced out of the market." The UMW's position, as stated in 1928 by John L. Lewis, is this: "Coal consumption does not depend on the price of coal to any great extent. Sales campaigns and price reductions do not and cannot boost coal consumption as in other industries. Considered as a whole, coal is a very minor item in the total cost of manufactured products." [53]

The union insists that the decline in demand for bituminous coal, relative to alternate fuels, is a result of shifting consumer tastes and has little to do with price. It believes, therefore, that remedial measures should concentrate on raising demand for coal, not on holding down or reducing prices. It argues, furthermore, that competition of other fuels is an invalid argument for holding down miners' wages. Even had buggy workers' wages been reduced to zero, the union argues, automobiles would still have replaced the buggy.

Union policy is, therefore, directed toward measures which will raise the demand for the product—and the demand for labor. UMW lobbyists work in the Congress of the United States for higher tariffs on oil imports and, at one time, minimum prices on domestic oil sales; for legislation which will conserve the scarce natural resource, natural gas; and against further public power projects, such as the St. Lawrence seaway.[54] Moreover, the union continually urges the coal operators to make their product more marketable by providing consumers with new techniques and devices which will assure them greater economy and convenience in the use of bituminous coal.

SUMMARY

The strategy of the UMW is to raise wage rates, to gain increased fringe benefits, to shorten hours of work, and to improve working conditions.

To achieve its purpose of raising the price of labor the UMW attempts to modify the conditions of both supply and demand for its "product." On the supply side union policy aims at a restriction of the quantity of labor available to the firms in the industry through seniority provisions in the collective agreement, policing of child la-

53. Statement to Senate Committee on Interstate Commerce, December 14, 1928, as reported in the UMW *Journal,* January 1, 1929, p. 5.

54. See, for example, "Joint Report of the International Officers," UMWA, *Proceedings of the 39th Convention,* p. 16.

bor laws, and similar techniques. On the demand side the UMW encourages substitution of capital for labor, thereby raising the marginal productivity of the miners; and it attempts to increase, or to prevent a further decline in, the demand for the product relative to competing fuels.

Regarded from a different angle, union policy is to reduce the number of operating mines and miners in the hope that this will result in a restriction in total output. If product demand can be kept at relatively high levels, the restriction in supply will result in higher selling prices. And higher product prices make possible the winning of higher wage rates.

CHAPTER 5

Political Aspects of Union-Management Relations

POWER RELATIONSHIPS IN THE INDUSTRY

THE COLLECTIVE bargaining process is a trial by combat. The process of negotiation is in essence an attempt by each side to impose its will on the other. Both parties constantly strive to enhance their bargaining power, i.e., "to increase the cost to the other party of *disagreeing* on the first party's terms relative to the cost of *agreeing* on the first party's terms." [1] If the cost to the coal operators of disagreeing on the UMW's terms is greater than the cost of agreeing with the union on the union's terms, then management's bargaining power is low relative to that of the UMW. It should be noted that this definition is framed in terms of both money and nonmoney costs.

A. H. Raskin has illustrated aptly the "power consciousness" of the miners' union:

> John L. Lewis worships power—his own power. The magniloquent president of the United Mine Workers loves to tell of a talk he once had with a West Virginia coal operator after an economic battle so fiercely fought that each man had come out with great respect for the other.
>
> In friendly fashion the union chief told the mine owner that he could not understand a man like him, endowed with brains, education, personality, fortune, "whose sole interest in life is the acquisition of wealth, wealth and more wealth."
>
> "Yes," was the operator's retort, "and I can't understand a man like you, with your great talent, whose sole interest in life is the acquisition of power, power and more power."
>
> When he repeats this story, Lewis rolls out the word "power" with a violence that sets ashtrays jumping. He laughs uproariously as he tells it.[2]

The union's awareness of the need for bargaining power is by no means a recent development. The earliest records of the UMW and

1. N. W. Chamberlain, *Collective Bargaining* (New York, McGraw-Hill, 1951), pp. 220–1.

2. A. H. Raskin, "Secrets of John L. Lewis' Great Power," *New York Times*, October 5, 1952, Magazine Section, p. 15.

its forebears indicate how vital to the achievement of union goals was superior bargaining strength. The official historian of the UMW reported:

Though [in 1873] we had what some considered strong trades unions, yet as a means to the end for which they were formed the ablest and most experienced considered them a failure. Doubtless those have accomplished much good in their own localities on minor questions. . . . But in almost every case, where there was determined opposition on the part of the giant corporations who now own and operate the American mines, they have had to give way as gossamer webs before the summer sun. If a mine was owned by a single individual, a district union might cope with him; but when a hundred men or more place millions of dollars in the hands of a board of three or four directors, their power is limited only by their will. No district or state association can cope and be successful against such a body single-handed.[3]

The preamble to the constitution of the National Federation of Miners and Laborers of the United States declared:

Local, district and state organizations have done much toward ameliorating the condition of our craft in the past, but today neither district nor state unions can regulate the markets to which their coal is shipped. . . . In a federation of all lodges and branches of miners' unions lies our only hope. Single-handed we can do nothing, but federated there is no power of wrong that we may not openly defy. . . .[4]

There can be no doubt that the union's leadership has learned well the lessons of the past. Union officials still seek to improve the power position of their organization. In several recent wage bargains the union won from the operators, along with some "worker-oriented" provisions, certain "union-oriented" concessions—conditions "intended to define the status of the union in the enterprise." [5] In 1939, for instance, the UMW traded an increase in wage rates for a union shop clause. In 1941 the union won a "most-favored-nation" clause, providing that should the union gain from any company terms better than those already in the national contract, the better terms would be applied to all companies. In 1947 the union gained the "willing and able" provision, which permits the UMW to call a work stoppage at any time; it achieved virtual management control over the miners' welfare fund; and it extracted from the operators an agreement to handle all controversies involving damage suits against the union,

3. C. Evans, *History of the United Mine Workers of America* (Indianapolis [UMWA? 1918?]), *1*, 46.

4. Ibid., p. 139. The constitution was adopted on September 12, 1885.

5. Ross, *Trade Union Wage Policy*, p. 23.

either through special machinery to be devised by the two groups or through collective bargaining.[6]

As is typical of collective agreements, that between the miners' union and the coal producers provides a formal statement as to the rights of management. The relevant section, entitled "Management of Mines," reads: "The management of the mine, the direction of the working force and the right to hire and discharge are vested exclusively in the Operator, and the United Mine Workers of America shall not abridge these rights."

Though this statement seems to preclude any infringement of the discretionary powers of management, the impact of union activities on those powers is considerable. Not only do union wage policies have an important effect on price-output decisions, including the nature of the production function of the firm, but the UMW also has gradually assumed a role in the day-to-day operation of the mines. Thus the collective agreement includes, in addition to those already noted, provisions for joint settlement of disputes, hiring and firing of employees in accordance with worker seniority, and the establishment of rules concerning safety conditions in the mine. It is this situation that constitutes what has been called rightfully "the union challenge to management control."

The bargaining power of any organization varies directly with the cohesion of its members to the avowed objectives of the group. It is one function of leadership to establish group discipline, to win the allegiance of the rank and file to the objectives set forth by those possessed of the discretionary powers. This was the first great chore of John L. Lewis when he became president of the UMW in 1919. At the time

> . . . the union was a union in name, but in fact was a group of large, individual domains loosely tied together in a national organization. Each mining district and its local chiefs guarded their own independence from invasion and domination by the national office. The miners summed up their conception of their rights in the term "autonomy." This included the right to elect their regional officials. While these local practices of autonomy were in keeping with the basic traditions of democracy, they were resulting in severe disunity within the union. "Autonomy" was developing into a fatal weakness. Each district would try to get the best deal it could, regardless of what happened to the miners in other parts of the country. . . . This

6. The third point was important because it freed the UMW from the strictures of the Taft-Hartley law and from the need for recourse to the National Labor Relations Board. At that time the UMW had no channel to the NLRB because the union's officers had refused to sign the non-Communist oath required by the Taft-Hartley Act.

practice was encouraged by the operators as it tended to set off one district union against the other and to breed discord and disunity within the union. The local officers seemed unaware of the disastrous effects of separate agreements, which undermined the whole idea of national collective bargaining.

For one district to get higher pay than an adjoining district simply meant that the lower wage rate prevailing outside would, in the last analysis, become the standard and shortly destroy the higher wage rate. The only hope for the miners was in a closely integrated national union which could deal with the coal industry on a nationwide basis.

Confronted with this situation, Lewis coldly decided to unite the union into a single disciplined force. The only way to do it was to develop a machine that could assure control over the entire union. . . . He was convinced from everything he knew and had seen in the union that the only way that this loosely-tied organization could be welded into a single, powerful force was to root out completely local autonomy and division of authority. It would mean taking over control of the union, lock, stock and barrel.[7]

The leadership of a group must have broad powers of discretion if there is to be continuity in the formulation and implementation of the association's objectives. As a practical matter, this means that power should be centralized, even though it is ultimately responsible to the members. This seems a fair characterization of internal political arrangements in the UMW. Wide discretionary powers are vested in the union president. All officers of the international union apparently owe their status to the president. And 70% of the presidents of district unions are in a similar situation, since they are "provisional" appointees.[8] The UMW, as a result, is run by an active minority

7. This account is taken from Lewis' friendly biographer, Saul Alinsky, *John L. Lewis* (New York, Putnam, 1949), pp. 37–8.

8. The power to appoint district officers is granted by Art. III, Sec. 2 of the Constitution of the UMW. The section reads in part: "Charters of Districts, Sub-Districts and Local Unions may be revoked by the International President, who shall have authority to create a provisional government for the subordinate branch whose charter has been revoked. This action of the International President shall be subject to review by the International Executive Board upon appeal by any officers deposed or any members affected thereby. . . . An appeal may be had from the decision of the Executive Board upon such order of revocation, to the next International Convention." This section is an excellent illustration of the attachment of American interest groups to the form but not the content of democracy, for the members of the International Executive Board are products of the Lewis "machine." What is more, the international officers maintain a firm grip on the convention machinery. The effectiveness of "provisionalism" from the point of view of the international officers is attested by a reliable report that by 1948 21 out of 31 UMW districts were being run by Lewis appointees. See *New York Times,* October 8, 1948, p. 21. For evidence of the persistent attempts by several locals to regain autonomy, see the lists of resolutions in the proceedings of every union convention from 1936 to 1946. All such resolutions were declared by the presiding officer to have been defeated by overwhelming majorities.

which controls policy objectives, tactical procedures, and all related matters.

Though the policy makers have broad powers of discretion, the cohesion within the union is derived in significant part from the fact that the miners believe that Lewis gets results. The following sarcastic comment by a member of the Congress of Industrial Organizations —then presided over by Lewis—is probably illustrative of the attitudes of most members of the UMW: "Boy, do I hate this man Lewis! We used to get 60 cents an hour. . . . My raise amounted to 80 cents a day or $16 a month, and out of that I pay $1 a month in dues. Do I hate this man John L. Lewis!" [9]

There is evidence, moreover, that where the miners encounter a conflict between the interests of their union and some other group to which they also owe allegiance, the conflict is resolved in the union's favor.

In May of 1943 the leadership of the United Mine Workers was threatening a strike that would, if it took place, almost certainly precipitate government seizure of the mines. The union members were asked whether they favored calling a strike, and they responded two to one in the negative. The primary reason given was the "desire not to let down the war effort, their confidence in President Roosevelt as a man who would give them a square deal." But in spite of this, more than three-fourths of the miners also stated that they would go out if the strike were called, because they "wouldn't let their union and their union leaders down." [10]

Group discipline among the coal operators is strikingly slack when compared with the monolithic structure of the miners' union. The bargaining power of the producers is, as a consequence, sharply diminished. As suggested in Chapter 2, there are important conflicts of interest between firms in the northern Appalachian fields and those in the southern Appalachians; between the commercial producers as a group and the captive operators; and between managers of low- and high-cost mines. These interest groups, it must be stressed, are not mutually exclusive. There are, for instance, captive mines in both the north and south. There are low-cost mines in the south and high-cost mines in the north. In a labor dispute involving a variety of issues

9. E. Levinson, *Labor on the March* (New York, Harper, 1938), p. 294, cited in Coleman, *Men and Coal*, p. 219.

10. M. Sherif and H. Cantril, *The Psychology of Ego-Involvements* (New York, John Wiley, 1947), p. 380, cited in D. B. Truman, *The Governmental Process* (New York, Knopf, 1951), pp. 160–1. The miners do not invariably resolve overlapping conflicts of interest in the union's favor, as notably attested by their rejection of Lewis' leadership in the 1940 national election. For other comment on this latter point, see S. Lubell, *The Future of American Politics* (New York, Harper, 1952), pp. 51–2.

an individual operator may find that his allegiances overlap, that his self-interest varies on separate issues from one faction to another. The coalition of operators is thus subject to dissolution at one or more points upon the application of strong pressure by the union—a situation upon which the UMW has capitalized heavily in the past.

The nature of the differences among the producers, and the bargaining advantage that redounds to the UMW as a result, are highlighted by analysis of some recent bargaining negotiations. The labor dispute in 1941, for instance, revealed dramatically the deep-seated antagonisms between commercial producers in the north and south.

The year 1941 marked the first time since the mid-twenties that more coal firms earned than lost money. Demand for coal was rising under the impetus of the defense program. Minimum pricing under the second Guffey Act had become a dead issue. The coal operators were clearly in no mood for a protracted work stoppage.

The UMW could not and did not fail to sense the changed temper within the industry. The decision was made to demand, along with a wage increase, the elimination of the 40¢ daily wage-rate differential between north and south. In this demand the union knew it could count on the support of the northern producers who had chafed under this discrimination for as long as the union had. The northern operators quickly agreed to the elimination of the differential. The southerners reacted violently, charging that this represented "a deal which the northern operators seek to make with the United Mine Workers in an effort to restrict tonnage." [11] So saying, the southerners bolted the Joint Conference and convened separately to form their own bargaining association.[12]

Charles O'Neill, spokesman for the northern group, tipped his hand when he sharply criticized the southern producers for their stand. "Despite assertions to the contrary . . . I charge that this conference is being sabotaged by several well-known southern coal operators in order that they may retain in its whole great amount the differentials in wages . . . that they now have in their favor as against northern coal producers." [13]

A vital sidelight to this question is the voting procedure of the Joint Appalachian Conference. All decisions are made under the "unit rule"; that is, the vote of the majority binds all those voting. Since the votes of the mine operators are weighted according to the tonnage they represent, the coalition of the captive and northern commercial operators has dominated the proceedings. It is clear, there-

11. *New York Times,* April 8, 1941, p. 16.
12. Ibid., April 12, 1941, p. 9.
13. Ibid.

fore, why the southern firms have insisted on separate negotiations with the union. At worst, they can accept the terms agreed to by the northern commercial and captive firms, and it is conceivable that at some time they may win terms more favorable than those in the industry bargain. In either case, moreover, the southern producers have as a last resort an appeal to public opinion, a technique which they used rather early in the 1941 dispute.[14] Ebersole Gaines, president of the newly formed Southern Coal Operators Wage Conference, declared that the northern operators had insisted that the "miners force the south to make concessions that are beyond the ability of the south to make. . . . I think the public will understand who is sabotaging the defense program. . . ." [15]

There was a reaction to the cleft in the operators' ranks, but what happened was not particularly what the southerners wanted to happen. The President rushed through Congress a two-year extension of the Guffey Act, "in the belief that its passage would hasten agreement between the operators and the United Mine Workers." [16] The Secretary of Labor, after first refusing to do so, certified the dispute to the National Defense Mediation Board.[17] And President Roosevelt, in an obvious attempt to mollify the southern operators, asked the Interstate Commerce Commission to study the freight-rate structure, described by the President as unfair to the south.[18]

Pending the decision of the NDMB, the union concluded a contract with both northern and southern producers, though the differential question was left unresolved. On the UMW's insistence, however, a most favored nation clause was inserted, under which the best terms obtained by the UMW from any firm would automatically apply to all firms. The clause, declared Lewis, would be used to per-

14. A union official has explained the unit rule to me in quite different terms. According to him, the unit rule requires an unanimous decision of all participants in the voting. But by this interpretation the southern firms would have a veto, and it would seem to be to their advantage to remain a part of the Joint Conference. To explain the secession of the southern firms the union hierarchy argues that the southerners are attempting to mobilize public opinion against the UMW by implying that the union is blackjacking the small firms into submission. It is true that the southern producers have been implacable foes of uniform wage rates and working conditions in the industry, and have sought in various ways to curb the UMW's power. But it is unreasonable to argue that the southerners, just to dramatize their weak bargaining position, would give up their power to block a settlement which operated to their distinct disadvantage.

15. *New York Times,* April 13, 1941, p. 33.

16. Ibid., p. 1.

17. Miss Perkins had at first refused to certify the case to the NDMB because she was well aware that John L. Lewis had made it known that the UMW would not cooperate with the Board. Ibid., April 18, 1941, p. 1. There is no indication in the record as to what caused Miss Perkins to change her mind, or Lewis to change his.

18. Ibid., April 24, 1941, p. 1.

mit the northern mines to reduce their wage rates by 40¢ per day if the southerners would not accept the elimination of the daily differential.[19]

The NDMB, as widely anticipated, recommended the elimination of the regional wage-rate differential. Now the southern firms were "outflanked" and within a few days they accepted the union's demands as "patriotic citizens" in the national emergency.[20]

In the years after World War II the demand for bituminous coal was somewhat below the wartime level. As a result, during much of the postwar period many commercial mines were able to fill all purchase orders of their customers by operating as few as three days a week. Under these conditions the commercial operators, northern and southern, high-cost and low-cost, had common cause in resisting union wage demands.

During the same years, however, the demand for steel and electric power remained at high levels. In those industries firms owning bituminous coal mines had a strong incentive to keep their mines in operation on a full work schedule. Moreover, in two or three of the postwar years there were lengthy strikes against major coal users, particularly steel-producing firms. Once those disputes had been ended, the steel companies were understandably anxious to prevent shutdowns in their coal mines.

In both 1946 and 1947 the UMW cracked the "solid front" erected by the bituminous operators, doing so by exploiting the known reluctance of the captive producers to endure prolonged work stoppages. John L. Lewis met privately with representatives of the captive mines, obtained their agreement to the union's demands, and then announced that the agreed terms would be applied to the rest of the industry. The low-cost commercial mines, particularly those in the northern Appalachians, now saw no further point in resisting the UMW. Counting on the union to apply the same terms to all coal firms, the low-cost commercial mines capitulated in short order, reckoning that their competitive position relative to other commercial mines would at worst remain unchanged and at best improve. The high-cost mines were left in an awkward position: if they rejected the union's demands, their mines would remain closed and their customers pre-empted by those operators who had accepted the union's terms. If the high-cost operators gave in to the UMW their total costs would be increased in equal amount with all other firms, but their average costs of production would rise relative to those of firms of greater productive

19. Ibid., May 21, 1941, p. 17.
20. Ibid., June 10, 1941, p. 17.

efficiency. By agreeing on the union's terms, then, the low-cost firms hoped to improve their competitive status relative to the marginal mines.

The lack of cohesion among the operators has effectively prevented the establishment of a durable unified front during negotiations with the union. There was between 1941 and 1946 a single employers' bargaining unit which represented all the Appalachian operators. But shortly after the war's end the southern producers withdrew from the Joint Conference and insisted that henceforth they would make separate agreements with the UMW. Since that time, collective bargaining has not been industry-wide in the strict sense, though the union's skillful use of its superior bargaining power has resulted in a uniform wage agreement for the entire Appalachian area. With or without a unified operators' bargaining group, therefore, the terms on which mine labor works have been the same for all employers.

The Union and the Government

While the UMW has pursued its aims largely by employing its considerable bargaining power against mine management, it has on several occasions sought the aid of government. This is not unusual; indeed, it seems to be an inevitable procedure for interest groups of all kinds. As David Truman has explained:

> The special characteristic of contemporary interest groups that has "inevitably" forced them to operate in the political sphere is this: unaided by the wider powers of some inclusive institutionalized group, they cannot achieve their objectives; interest groups of the association type cannot, without . . . mediation, perform their basic function, namely, the establishment and maintenance of an equilibrium in the relationships of their members. . . . Such groups supplement their own resources by operating upon or through that institutionalized group whose powers are most inclusive in that time and place. With . . . local exceptions . . . that institution today is government.[21]

Whenever the UMW has felt that its own power was sufficient to realize its policy objectives, it has adopted an uncompromising attitude of "laissez-faire." When, on the other hand, the union has found itself unable "to establish and maintain an equilibrium in the relationships" of the employers and the miners, it has demanded public assistance. These considerations account for the apparent vacillation of the UMW on the issue of governmental intervention in the affairs of the bituminous industry over the past 35 years.

In 1917 the union declared publicly that it saw no need for public

21. *The Governmental Process*, pp. 105–6.

intervention in the coal industry.[22] This was a period during which demand for labor was high, wage rates were rising, and union power had reached an unprecedented peak. Within two years, in contrast, demand for the product had receded to more "normal" levels. In April 1919 the union policy committee "adopted a sound economic program, which provided . . . for government regulation of the bituminous coal industry. . . ." [23]

In 1933, when the National Industrial Recovery Act was before the Congress, the UMW gave its enthusiastic backing. "It is a matter of particular pride to the officers [of the union] . . . that a number of the important principles of the Recovery Act were first suggested to the Federal Congress and to the nation by the United Mine Workers of America." [24] But 15 years later the relationship between the union and the operators had been restored to a condition satisfactory to the UMW. Asked in 1947 if he favored government operation of the mines, John L. Lewis replied: "No, perish the thought. . . . I favor free enterprise." [25]

What is unique about the interaction of the UMW and the Federal Government is that the latter has often collaborated with the union rather than acting as a mediator between the conflicting private associations. The government, willingly or unwillingly, has on several occasions permitted itself to be maneuvered by the UMW into a posture which has amounted to outright support of the union's demands upon the employers. A somewhat superficial indication of the dilemma in which the executive branch became involved is revealed by an exchange between President Roosevelt and Secretary of the Interior Ickes during the bituminous strike of 1943. Mr. Ickes is reported to have asked the President: "Do you want me to mine coal or smash Lewis?" The answer was "Mine coal." [26]

The techniques by which the miners' union has involved the Federal Government in the union's efforts to maximize its bargaining power are typified by the events of 1941. In the contract signed in that year by all commercial producers the union had won a provision which allowed it to order a work stoppage at any time in order to "maintain competitive parity" in the industry. This protective clause was a fur-

22. C. L. Sulzberger, *Sit Down with John L. Lewis* (New York, Random House, 1938), p. 36.

23. UMWA, *Proceedings of the 33rd Convention* (1934), p. 25.

24. Ibid., p. 26. There seem to be others who claim the "honor" of originating the underlying principles of the NRA. See, for example, M. Josephson, *Sidney Hillman. Statesman of American Labor* (New York, Doubleday, 1952), p. 362.

25. C. A. Madison, *American Labor Leaders* (New York, Harper, 1950), p. 197.

26. Dispatch from Arthur Krock to *New York Times,* November 14, 1946, p. 28. The story may be apocryphal, but it certainly illustrates the problem.

ther safeguard for the union's next move—which was not long in coming.

On September 14, 1941 the mineworkers struck the captives. The sole issue was the union shop, which the UMW had won two years before from the commercial operators. As in 1939, the operators of the captive mines—especially the steelmasters—refused to yield, knowing that by doing so they would then have to make the same concession to the Steel Workers' Organizing Committee (later the United Steel Workers of America). But the union held most of the high cards in this game. The Federal Government's growing concern over the international situation would force it to side with the union. The commercial operators had nothing to gain from standing by the captives; if the union invoked the protective clause, the commercial mines would be shut down again just when the coal trade was brisk. Finally, the union had the most-favored-nation provision in the Appalachian contract as a weapon in any legal showdown.

The National Defense Mediation Board, fearful of setting a precedent which would apply in the steel and shipbuilding industries, refused at first to rule on the union shop issue.[27] The UMW, which had ended its strike pending an NDMB decision, pulled the miners out of the mines again on October 26. Thereupon the mediation board agreed to review the case and to rule on the union shop issue. In return, Lewis ordered a truce in the strike and his miners once more trooped back to work.[28]

On November 10 the mediation board announced its opposition to the union shop in the coal mines, representatives of the American Federation of Labor voting with management members of the board. Only the CIO representatives voted in favor.[29] The UMW promptly called a new strike for November 17. And on November 19 workers in the commercial mines began to walk out in sympathy. Union power was thus brought firmly to bear on President Roosevelt. The latter hesitated for a few days, apparently in the hope that public opinion would either force Lewis to back down or impel the miners to return to work without an order from Lewis. Neither Lewis nor his men gave way; Mr. Roosevelt did. He appointed a special arbitration board—effectively evading the NDMB—to rule on the issue.

Though the special panel seemed to be constituted equitably, it must

27. *New York Times,* October 25, 1941, p. 1.
28. Ibid., October 30, 1941, p. 1.
29. The AFL justified its stand on its historic opposition to a union shop which was imposed on industry by "government decree." *New York Times,* November 11, 1941, p. 1. More likely, the AFL members were delighted to have this opportunity to retard the growth of the "upstart" CIO.

have been clear to all that the union was going to get what it wanted. The board had three members: Dr. John R. Steelman, special assistant to the president and the "public member"; Benjamin Fairless, president of U. S. Steel, the "industry member"; and John L. Lewis. On December 7, 1941 the announcement was made that the panel had voted 2-1, Mr. Fairless opposing, to award the union shop.[30]

While the United Mine Workers is not a political interest group in the sense, say, of the American Medical Association or the American Legion, the union's frequent need for governmental support has led it into political action. In part, the UMW's political efforts have been motivated by general considerations; Lewis' ardent support of President Roosevelt in the 1936 national election rested on his belief that "labor has gained more under President Roosevelt than under any President in memory. Obviously it is the duty of labor to support Roosevelt 100 percent." [31] Implicit in this declaration was Lewis' recognition that there were many things yet to be done in labor's behalf, specifically the enforcement of the National Labor Relations (Wagner) Act and the enactment of legislation to control wages and hours.

In larger part, however, the UMW's sallies into politics demonstrate the validity of the proposition that "power of any kind cannot be reached by a political interest group, or its leaders, without access to one or more key points of decision in the government." [32] It is this consideration which best explains acts such as the UMW's contribution of $500,000 to the 1936 campaign fund of the Democratic party. As Lewis himself put it: "Everybody says I want my pound of flesh, that I gave Mr. Roosevelt $500,000 for his 1936 campaign, and I want quid pro quo. The United Mine Workers and the CIO have paid cash on the barrel for every piece of legislation that we have gotten. . . . Is anyone fool enough to believe for one instant that we gave this money to Roosevelt because we were spellbound by his voice?" [33] And again, after the President's outburst of "a plague o' both your houses" following the bloody steel strike in 1937, Lewis declaimed: "It ill behooves one who has supped at labor's table and who has been sheltered in labor's house to curse with equal fervor and fine impartiality both labor and its adversaries when they become locked in deadly embrace." [34]

The relationships between Lewis and the White House have occupied most of the public's attention. It is less widely appreciated that

30. *New York Times,* December 8, 1941, p. 1.
31. Quoted in Josephson, *Sidney Hillman,* p. 395.
32. Truman, *The Governmental Process,* p. 264.
33. Alinsky, *John L. Lewis,* p. 177.
34. Speech on Labor Day, 1937, cited in Coleman, *Men and Coal,* p. 175.

the union president has made strenuous efforts to gain access to the Bureau of Mines, another key point of decision for the bituminous industry. As the federal government's center for the collection of technical and economic data on the industry, the Bureau has undoubted influence over any legislation presented to Congress affecting coal mines. Indicative of the latent powers in the Bureau is the fact that in the later years of the federal minimum pricing program the functions of the National Bituminous Coal Commission were transferred to the Bituminous Coal Division of the Bureau. It is the Bureau of Mines, too, which is responsible for enforcing the Federal Mine Safety Code.[35]

The bitter battles of the past between Lewis and Secretaries of the Interior Harold Ickes and Julius Krug take on real meaning in view of the multifarious activities of the Bureau of Mines. The same considerations explain, too, the recent unsuccessful attempt to prevent the appointment of James Boyd as Director of the Bureau. Either an official is friendly to the union and its objectives or he is subjected to an attack waged with fervor and a matchless vocabulary.

In most of his political activities Lewis has enjoyed the backing of his constituents in the union, though it is likely that the bulk of the membership has been passive rather than active in its adherence. As noted previously, the UMW is characterized by genuine cohesiveness, based largely on the unwavering confidence of the members—"Ole John L. may be rough, but he's always working for us." [36] But cohesion in the union is not invariant and does not obtain on all issues, as the famous case of the 1940 national election attested. Defections of that kind are essentially to be explained by the fact that "labor unions as organizations and as memberships are oriented only in the economic sphere . . . If a labor leader begins to work for a political act not immediately and obviously involving an economic gain for his membership, immediately he will be accused of working for personal glory and selfish power." [37]

This reasoning is not completely satisfactory, for the miners apparently followed Lewis' lead in the elections of 1936 and 1952:

35. It need scarcely be said that safety conditions in the mines are of paramount importance to the miners. Lewis has apparently recognized an even more important aspect of safety regulations. The *New York Times* of April 12, 1947 (p. 1) reported that Lewis was preparing to ask for legislation which would give the UMW the power to shut down any mine considered by union-paid inspectors to be unsafe. The potentialities of such a law for enhancing the UMW's power to punish recalcitrant operators are immense. There is no evidence to indicate that the matter has been pressed, which suggests that the union's announcement was primarily a tactical maneuver designed to promote its current wage demands.

36. Quoted in C. Wright Mills, *The New Men of Power* (New York, Harcourt, Brace, 1948), p. 66.

37. Ibid., p. 236.

that is, they voted Democratic.[38] In other words, the miners have balloted according to their personal political biases. As Samuel Lubell has argued, the New Deal gave effective political representation for the first time to the lower income elements, whatever their race, religion, or cultural differences.[39] Since the coal-mining population is a mélange of all these elements—partially assimilated immigrants, Negroes, Catholics, and so on—it is understandable that their vote should be consistently Democratic.

The Union as Entrepreneur

The preceding section indicates that the union is an important locus of initiative in the bituminous industry. The motive power for change in the structure and behavior of the industry has originated almost completely with the United Mine Workers. Only rarely have the producers, singly or together, acted independently of the union in an effort to alter an existing arrangement.

There have been several evidences of the union's adoption of the entrepreneurial function. The organization of the Central Competitive Field after 1898 was an outgrowth of a strike for union recognition. Bargaining relationships thereafter functioned on an interstate rather than a local basis. Similarly, the regularization of competition based on a uniform structure of wage rates was made possible by the union's struggle for industry-wide bargaining during the 1930's. Union action was primarily responsible, too, for the enactment of the bituminous coal stabilization (Guffey) acts, which established the principle of legal price minima for coal sellers. And, as will be pointed out in later chapters, union policies have altered the patterns of competition among coal-producing regions.

The central role of the union bears on the interesting generalization that labor organizations have developed as a response to the stimulus of monopsonistic buyers of labor. In a recent exposition of this hypothesis the bituminous industry was explicitly excluded, on the grounds that the strong labor organization preceded the development of a single employers' bargaining unit.[40] The validity of the generalization for all industries may well be in doubt, though this is not the place to test it. It does apply to the bituminous industry, despite the fact that Professor Galbraith excluded coal mining. In this industry it is true that there has not yet developed an employers' association capable of counterbalancing the power of the UMW. But it

38. During the campaigns of 1944 and 1948 Lewis' official position was neutral, though it is generally agreed that he has been a life-long Republican.

39. Lubell, *Future of American Politics,* pp. 28 ff.

40. J. K. Galbraith, *American Capitalism. The Concept of Countervailing Power* (Boston, Houghton Mifflin, 1952), pp. 121–3.

was not always thus, as the above excerpts from the official history of miners' organizations testify eloquently for the years before 1890. Similarly, the full power of the UMW was insufficient to carry out the organization of the southern Appalachian fields during the twenties.

That a union should arrogate to itself the responsibility for the stabilization of a highly competitive industry is unusual, though not unprecedented. The stabilization activities of the garment and clothing workers' unions are well known. The notable aspect of the UMW policy as compared with that of the International Ladies' Garment Workers Union, for example, is that the coal miners are imposing their wills on a reluctant industry, quite in contrast to the cooperative arrangement in the garment industry. Because of this difference, the considerable success of the UMW is the more remarkable.

CHAPTER 6

The Union's Effects on the Wage Level

THE MOVEMENT OF WAGE RATES

IN 1933 wage rates in bituminous coal mining began an upward secular trend. The first Appalachian contract (October 1933) established a daily wage of $5.00 in the northern fields, $4.60 in the southern.[1] By 1948 all firms, north and south, were paying daily wage rates of $14.05.[2]

The union also extracted from the coal operators some important fringe benefits. Premium rates for overtime work were instituted in 1937. Vacation payments, which began at $20 per man per year in 1941, had risen to $100. The operators in 1943 assumed the expense of providing tools, miners' lamps, and smithing charges. Compensation for travel time (portal-to-portal) was won in the same year, as was a paid lunch period (later increased from 15 to 30 minutes per day). Shift differentials became part of the national contract in 1945. A noncontributory welfare fund for the miners, completely financed by per ton royalty payments by the operators, was begun in 1946. Royalty payments rose from an initial rate of 5¢ per ton in 1946 to 20¢ per ton in 1948.[3]

One analyst has computed the value of monetary concessions granted the UMW by the southern producers between 1941 and 1948 as amounting to $10.06 per day. He found that in July 1948 miners were being paid $15.66 per day for the same number of working hours as in January 1941. This amounted to a percentage increase in total compensation of nearly 180%.[4]

1. The custom in the industry is to use the basic inside workers' day rate as the average for all workers. Rates for outside workers are somewhat lower than for men employed inside the mine. Since 1933, too, tonnage rates have risen. Little attention will be given to piece rates since 1. they have become less important each passing year; and 2. piece rates after 1933 were adjusted in successive contracts so that daily earnings closely approximated those of inside day workers.

2. Hearings, Senate Committee on Banking and Currency, *Economic Power of Labor Organizations,* Pt. 1, p. 310.

3. Ibid., pp. 307–11. The impact of the welfare fund on labor mobility and total mine employment is discussed in Chap. 7.

4. J. Backman, *Bituminous Coal Wages, Profits, and Productivity* ([Washington?] Southern Coal Producers' Association, 1950), p. 17.

The over-all changes can be appreciated more readily by comparing the movements of average hourly, weekly, and annual earnings. Between 1935 and 1948 average hourly earnings rose from 75¢ to $1.90, an increase of 153%. Average weekly earnings increased by an even greater amount, climbing from less than $20 in 1935 to $72 in 1948, a gain of over 270%. To a large extent the greater relative increase in average weekly as compared with average hourly earnings was attributable to a 43% rise in the average weekly hours of labor. The longer work week likewise contributed to the more than proportionate increase in average annual earnings, which rose by more than 250% between 1935 and 1948. Nonetheless, average weekly hours began to decline after 1944, but earnings kept rising.

Table 7. Hours and Earnings in the Bituminous Industry, 1935–48

Year	*Average Weekly Hours*	*Average Hourly Earnings*	*Average Weekly Earnings*	*Average Annual Earnings*
1935	26.4	$0.745	$19.58	$ 957
1936	28.8	.794	22.71	1103
1937	27.9	.856	23.84	1170
1938	23.5	.878	20.80	1050
1939	27.1	.886	23.88	1197
1940	28.1	.883	24.71	1235
1941	31.1	.993	30.86	1500
1942	32.9	1.059	35.02	1715
1943	36.6	1.139	41.62	2115
1944	43.4	1.186	51.27	2535
1945	42.3	1.240	52.25	2629
1946	41.6	1.401	58.03	2726
1947	40.6	1.633	66.86	3201
1948	37.7	1.897	72.06	3387

Source: U. S. Bureau of Labor Statistics, compiled in National Coal Association, *Bituminous Coal Data, 1935–1948* (Washington, 1949), p. 101, and *Bituminous Coal Annual 1949* (Washington, 1949), p. 70.

Did the mineworkers keep pace with the wage gains realized elsewhere in the economy? During 1935–48 the miners' average hourly earnings rose by $1.15 or 153%. The increase for iron and steel workers during the same period was 85¢ per hour or 139%; for automobile workers, 88¢ per hour or 119%; and for all manufacturing workers, 78¢ per hour or 140%.

The relatively rapid rise in the wages of bituminous miners is reflected in the movement of average annual earnings.[5] The rise for the mineworkers during 1935–48 amounted to 253%; for iron and steel workers, 162%; for all manufacturing industry, 150%; and for all industries, 144%. In absolute terms the average annual earnings of coal miners in 1935 were $259 *less* than the average of all workers in manufacturing industry; while in 1948 miners were earning $347 *more* per year than employees of manufacturing industry.

Table 8. Average Hourly Earnings in All Manufacturing and in Selected Industry Groups, 1935–48

Year	*Bituminous Coal*	*All Manufacturing*	*Iron and Steel*	*Automobiles*
1935	$0.745	$0.550	$0.612	$0.739
1936	.794	.556	.620	.774
1937	.856	.624	.737	.891
1938	.878	.627	.753	.925
1939	.886	.633	.739	.929
1940	.883	.661	.755	.948
1941	.993	.729	.833	1.042
1942	1.059	.853	.939	1.169
1943	1.139	.961	1.033	1.234
1944	1.186	1.019	1.082	1.270
1945	1.240	1.023	1.100	1.256
1946	1.401	1.084	1.195	1.333
1947	1.633	1.221	1.343	1.473
1948	1.897	1.327	1.464	1.616
Increase: 1935–48				
Absolute	1.152	.777	.852	.877
Percentage	153.0	140.0	139.0	119.0

Source: U. S. Bureau of Labor Statistics, compiled in National Coal Association, *Bituminous Coal Data, 1935–1948,* pp. 101, 107

Comparison of the gains of miners relative to other workers must allow for the comparatively low level of miners' wages at the start of the period under review. Even so, it is clear that the level of wage rates in bituminous coal mining has risen more rapidly than in other important sectors of the economy. This raises a significant question: Had the miners not been unionized, would they have realized the same or similar wage gains?

5. It should be kept in mind that average annual earnings are a result of the interplay of adjustments in wage rates, in productivity, in the factor-mix, and in the level of employment in the industry. Changes in average annual earnings are only partially a result of changes in wage rates.

Table 9. Annual Earnings per Full-time Employee, Bituminous Coal, All Manufacturing, and Iron and Steel and Their Products, 1935–48

Year	*Bituminous Coal*	*Manu-facturing*	*Iron and Steel*	*All Industries*	*Spread: Coal and Manufacturing*
1935	$ 957	$1216	$1295	$1153	−259
1936	1103	1287	1446	1199	−184
1937	1170	1376	1591	1270	−206
1938	1050	1296	1359	1238	−246
1939	1197	1363	1549	1269	−166
1940	1235	1432	1643	1306	−197
1941	1500	1653	1923	1450	−153
1942	1715	2023	2283	1719	−308
1943	2115	2349	2637	1964	−234
1944	2535	2517	2781	2120	+ 18
1945	2629	2517	2792	2207	+112
1946	2724	2517	2696	2369	+207
1947	3213	2793	3063	2595	+420
1948	3387	3040	3392	2809	+347

Source: U. S. Department of Commerce, compiled in J. Backman, *Bituminous Coal Wages, Profits and Productivity* ([Washington?], Southern Coal Producers' Association, 1950), p. 29

Several writers have dealt with this problem, though each has generalized about the whole economy rather than analyzing changes in individual industries.[6] An imposing list of factors contributing to alterations in wage rates can be compiled from these studies: changes in the level of employment in each industry; an oligopolistic market structure; changes in productivity; changes in output; the proportion of labor cost to total cost; competitive conditions in the product market; and the changing skill and occupational content of the industry. There appears to be general agreement, however, that "unionization has been a necessary, but not sufficient condition for larger-than-average increases in earnings." [7]

No matter how the basic data are handled it is apparent that during

6. See, for example, J. T. Dunlop, "Productivity and the Wage Structure," in *Income, Employment and Public Policy. Essays in Honor of Alvin Hansen* (New York, Norton, 1948), pp. 341–62; Ross, *Trade Union Wage Policy,* pp. 113–33; A. M. Ross and W. Goldner, "The Interindustry Wage Structure. Forces Affecting the Interindustry Wage Structure," in *Quarterly Journal of Economics, 64* (May 1950), 254; and J. W. Garbarino, "A Theory of Interindustry Wage Structure Variation," in ibid., 282.

7. Ross and Goldner, "Interindustry Wage Structure," p. 281.

the period under review there was a more than average increase in the money wages of bituminous miners. One tabulation indicates, for example, that during 1933–46 straight-time hourly earnings of the miners increased 89.4¢, almost 30¢ above the average for all industries and by far the largest occurring in any single industry.[8]

What is most striking is that none of the several factors allegedly influencing earnings along with unionism operated in bituminous coal mining. Ross and Goldner conclude that rising employment and an oligopolistic market structure are the additional forces which help unionism to influence earnings. Employment actually declined in the bituminous industry during 1933–46. Moreover, only in a formal sense could the bituminous industry be dubbed oligopolistic—and then for but four years in the whole period considered.[9]

Neither do the conditions postulated by Garbarino apply to bituminous coal. The industry was not characterized by a high degree of output concentration: there were some very large firms and financially affiliated groups, but these had a relatively minor influence on price policy in the industry. Furthermore, there is no clear evidence that changes in productivity accounted for the wage increases in the industry. Data already introduced [10] indicate that average output per man-day was rising during most of 1933–46. Though there are no data on marginal productivity, it can be assumed that the marginal productivity schedule shifted to the right. This follows from the fact that during 1933–46 there was an increasing use of capital equipment in the industry, as well as a steady improvement in the technical efficiency of the machinery in use. To the extent that marginal productivity underlies the demand for productive services, the effect of capital accumulation was to raise wage rates. Just how much rates were raised is impossible to say, since there were counterbalancing influences constantly at work. From 1933 to 1940 the supply curve of labor was highly elastic—there was a sizable pool of unemployed miners; while from 1941 to 1946 it was both inelastic and shifting to the left—alternative job opportunities, both military and civilian, sharply reduced the number of available mine employees.

Dunlop's suggestion that changes in hourly earnings are related to changes in output does not explain the relatively greater gains of the

8. Ibid., p. 276.

9. Ross concludes that bituminous coal was made imperfectly competitive "by virtue of the Guffey Act." Ross, *Trade Union Wage Policy,* p. 126. The Guffey Act of 1937 (the second Bituminous Coal Conservation Act) operated fitfully at best, officially expiring in 1943. The principal stabilizing agency was in reality the UMW, to the extent that the union prevented wage cutting as an instrument for price competition. Even so, there was constant "price chiseling" in the industry from the beginning of public intervention in 1933.

10. See Table 6 in Chap. 3.

miners. Output of coal—and thus the (derived) demand for labor—did increase considerably during the period 1933–46. But the demand for coal is itself a derived demand, the principal coal consumers being the steel mills, the electric utilities, and the railroads. Average straight-time hourly earnings in the bituminous industry rose in greater amount between 1933 and 1946 than in any of those other industries.[11]

Dunlop has held, too, that "wage and salary rates would be expected to increase most . . . where labor costs are a small percentage of total costs [and] where the enterprises are in strong bargaining power with the purchasers of their output. . . ." [12] The assumption underlying the first proposition is that employer resistance to changes in wage rates tends to be weaker when labor cost is small relative to total cost. It is well known, however, that labor cost in coal mining averages 60% of total cost. Moreover, no case can be made that coal sellers had strong bargaining power vis à vis their customers except during the war years. From 1933 to 1941 the competitive position of the industry was such as to require almost continuous public intervention. During World War II the bargaining power of coal producers was much greater than previously, because of the high level of demand together with a decline in the availability of competing fuels. This may have had some influence on the movement of wage rates during the war, but such changes in rates as did take place in the coal industry were minor.

None of the propositions which have been offered to explain variations in the interindustry wage structure appears applicable to coal mining. No other conclusion seems possible than that the greater gains of coal miners compared to other workers, union and non-union, were directly attributable to the superior bargaining power and aggressiveness of the United Mine Workers. Is such a generalization warranted without some qualification? It would be if we could assume that the individual miner's supply curve of labor was of no importance in the determination of wage rates. There can be no certainty on this point.

As is discussed more fully in Chapter 7, the size of the mine labor force has been shrinking steadily since 1940. There can be no doubt that the continuing high levels of income and employment in the economy at large have opened up new horizons for miners and their families, encouraging them to leave the unpleasant and dangerous coal industry for other kinds of work. In this light the rising level of wage rates in coal mining may be explained as a reaction to the steady attrition of the mine labor force. To put it another way, wage rates in the

11. See Table 7-2 in appendix.
12. Dunlop, "Productivity and the Wage Structure," pp. 360–1.

industry had to be high enough to elicit the required amount of labor. There is no way, unfortunately, to determine whether at any time the same number of miners would have worked at some wage rate lower than that prevailing.

It is of some interest, however, that average hourly earnings in the coal industry rose far less spectacularly before 1940 than after. The gains of miners between 1933 and 1940 were somewhat less in relative terms than workers in each of the industry groups listed in Table 8. Average hourly earnings of coal miners in this period of substantial unemployment throughout the economy rose 19%; of workers in all manufacturing industry, 20%; of iron and steel workers, 23%; and of automobile workers, 28%. In absolute terms the average hourly earnings of miners increased by 14¢; of workers in all manufacturing industry, 11¢; of iron and steel workers, 14¢; and of automobile workers, 28¢.

Though for obvious reasons these figures cannot be pushed too far, they do suggest that aggressive unionism is less important as a device for raising wage rates than the mere existence of plentiful job opportunities in the economy at large. This point is emphasized by the fact that the miners' union during the thirties was a considerably stronger organization than those in the steel and automobile industries.

At the same time, the impact of the UMW on the level of wage rates cannot be disparaged. Since working conditions in the iron-mining, copper-mining, and lead- and zinc-mining industries are analogous to those in bituminous coal, it is logical to suppose that similar influences were at work in each of those industries as far as the level of wage rates and employment is concerned. Specifically, if it is argued that wage rates in bituminous mining were driven up because coal miners sought less hazardous employment, the same argument should be applied to the behavior of workers in the other mining industries. During 1939–50 average hourly earnings in the mining industries behaved as follows: [13]

			Increase	
	1939	*1950*	*Absolute*	*Per Cent*
Iron Mining	73.8 cents	$1.515	.777	103
Copper Mining	67.9	1.601	.922	136
Lead and Zinc Mining	68.3	1.565	.882	130
Bituminous Coal Mining	88.6	2.010	1.124	128

13. Data from U. S. Department of Labor, *Monthly Labor Review* (June 1950), p. 687; (June 1951), p. 741.

In the same period total employment in the four industries changed as follows:[14]

	1939		1950	Increase Absolute	Increase Per Cent
Iron Mining	21.1	thousand	31.9	18.8	89
Copper Mining	25.0		24.8	—0.2	—0.8
Lead and Zinc Mining	16.3		17.2	0.9	5.5
Bituminous Coal Mining	372.0		351.0	—21.0	—5.6

Finally, these were the indexes of changes in output per man-hour:[15]

	1939	1950	Increase
Iron Mining	100	125.4	25.4
Copper Mining	100	160.8	60.8
Lead and Zinc Mining	100	86.1 *	—13.9
Bituminous Coal Mining	100	123.3	23.3

* Figure for 1949.

These statistics do not prove that the UMW succeeded in pushing wage rates above the level which market forces would have established. That is clear merely because of such atypical developments as the decline in the index of output per man-hour in lead and zinc mining. The data suggest, however, that it was only in the bituminous industry that rising wage rates were associated with a significant reduction in employment. Moreover, apart from lead and zinc mining, output per man-hour rose least in bituminous coal. To the extent that the figures above are valid indicators of the forces at work, the United Mine Workers was apparently more an active than a passive agent in the acknowledged increase in miners' wage rates during 1939–50.

Cyclical Fluctuations in Wage Rates

The UMW, like many another trade union, has long sought to protect wage rates against reductions brought on by downward fluctuations in demand for labor. There are strong motives for putting floors under wage rates: reductions wipe out previous gains and must laboriously be recouped at a later time; cuts in wage rates tend to alienate the membership and weaken the solidarity of the union.

14. Data for 1939 from U. S. Department of Commerce, *Statistical Abstract of the United States, 1949* (Washington, 1951), Table 223, p. 206; and for 1950, U. S. Department of Labor, *Monthly Labor Review* (May 1951), p. 592.

15. Data from U. S. Department of Labor, *Productivity Trends in Selected Industries,* Bulletin 1046 (1951), pp. 19–25.

Prior to 1933 the miners' union was virtually powerless to check declines in wage rates during cyclical fluctuations. During the general depression of 1893–97, for example, average hourly earnings declined from a level of 18.8¢ to 13.8¢.[16] Similarly, average hourly earnings fell from 64.6¢ in 1929 to 41.7¢ in 1933.[17] Since 1933, the UMW has taken no reductions in wage rates, no matter what the prevailing level of demand for labor.[18] During the general recession, beginning in 1937, earnings increased from 85.6¢ to 88.6¢ in 1939.[19] And in the recession which began in late 1948, average hourly earnings rose from $1.90 to $2.01 in 1950.[20] In fact, throughout the entire period 1933–50 average hourly earnings never once fell. They rose, on the contrary, in every year but 1940, when they were unchanged from the level prevailing in 1939.

These findings may suggest that the price paid by the union for downward inflexibility of wage rates was increased unemployment during periods of falling demand for labor. The data do, in fact, indicate that the number of employed fell from 492,000 in 1937 to 422,000 in 1939. This constitutes no proof that employment would have remained unchanged had miners offered their services at lower wage rates. Between 1929 and 1933 wage rates were quite flexible, yet total men employed diminished from 503,000 to 419,000. What this means, of course, is that the demand for labor is not a simple function of the level of wage rates. It is affected by several variables which are of at least equal importance to wage rates. These other variables and their relationship to the decline in men employed after 1937 will be explored in a later chapter.[21]

Summary

The UMW's strategy is to raise the general wage level "like cutting a dog's ears—a little at a time, so it won't hurt so much." This plan of action has been unusually successful, in that miners' wages

16. P. H. Douglas, *Real Wages in the United States* (Boston, Houghton Mifflin, 1930), p. 152.

17. U. S. Bureau of Labor Statistics, cited in Berquist and Associates, *Economic Survey*, p. 296.

18. The nearest thing to a wage-rate reduction countenanced by the union was its 1939 offer (accepted by the operators) to extend the wage terms of the 1937 contract without change for two more years. In return, however, the UMW won the union shop from the commercial operators and the elimination of the penalty clause from its contract with the captive producers.

19. U. S. Department of Labor, as reported in National Coal Association, *Bituminous Coal Annual, 1950* (Washington, 1950), p. 12.

20. U. S. Department of Commerce, *Statistical Abstract of the United States, 1951*, p. 201. Available data indicate the same "stickiness" of wage rates elsewhere in the economy, particularly in unionized industries.

21. See Chap. 7.

have risen much more rapidly than in other industries, union or non-union. Moreover, the UMW has succeeded in putting a floor under wage rates, so that the wage structure has become impervious to fluctuations in product demand. In the union view, "wages have been taken out of competition"—meaning both that miners do not compete with one another wage-wise and that coal producers cannot use wage rates as a lever for competitive price reductions.

CHAPTER 7

The Union and Interregional Competition

Elimination of Regional Wage-rate Differentials

Because labor cost is so large a proportion of total production cost in bituminous mining, the wage policy of the UMW has an important influence on the industry. Firms must operate usually on profit margins of 2¢ or 3¢ per ton. Small differentials in wage rates among competing producers will, as a result, affect significantly the competitive statuses of the firms.

The UMW has a profound interest in a uniform and stable structure of wage rates throughout the industry. Uniformity of wage rates, it must be noted, is a precondition of stability when demand for labor is declining. The unhappy experiences during the twenties impressed upon the union leadership that regional wage differentials which bestow a cost advantage upon one region expose all miners to competitive wage-rate reductions. Put another way, unless demand for labor is stable or rising, price competition among sellers of the product will open the way for demands in the high-wage areas that wage rates be lowered to "competitive" levels, i.e., to the level of rates in the low-paying area. Uniformity of the wage-rate structure is thus a necessary but not sufficient condition for the stability of wage rates in the industry.

In the Appalachian coal region prior to 1933 wage rates in the southern fields were considerably less than those in the north.[1] While the amount of the north-south differential was reduced under the NRA coal code, daily wage rates in the south after 1933 were still 40¢ less than in the north.[2] This 40¢ differential withstood repeated union

1. In May 1933, just before the coal code went into effect, the modal rate for trackmen in western Pennsylvania ranged between $3.00 and $3.25 per day. In contrast, mines in Southern Numbered 2 were paying a modal rate of between $2.25 and $2.50 per day. Berquist and Associates, *Economic Survey*, p. 283.

2. Because the mines in northern West Virginia were geographically in the south, but economically in the north, a compromise in the code provided that rates there would be 24¢ per day less than the northern scale. In the collective agreement of 1934, wage rates were increased by 40¢ per day in all Appalachian coal fields except northern West Virginia, where rates were raised by 64¢ per day. Ibid., pp. 318 ff.

attacks until 1941. As already stated, there was in 1941 a brief but bitter struggle culminating in union victory—a uniform structure of wage rates for the entire Appalachian region.

Though southern producers professed deep concern for their ability to survive without the differentials provided in the 1933 contract, cost data indicate that the immediate impact of the 1933 changes in wage rates on competitive cost positions was negligible.[3]

Table 10. Average Labor Costs before and after the NRA Coal Code, 1933

Producing District	*May 1933 Per Ton*	*Dec. 1933 Per Ton*	*Increase Due to Code Per Ton*	*Increase Due to Code Per Cent*
E. Penn. (1)	$.81	$1.07	$.26	32.6
Md. and Upper Potomac (1)	.71	1.14	.43	61.3
W. Penn. (2)	.63	1.01	.38	60.5
No. W. Va. (3)	.48	.78	.30	60.7
Ohio (4)	.65	.96	.31	46.6
Panhandle of W. Va. (6)	—	.97	—	—
So. No. 1 (7)	—	.96	—	—
So. No. 2 (8)	.56	.89	.33	58.3

Note: Numerals in parentheses represent the designations given those districts under the Bituminous Coal Conservation Act of 1937 and the regulations of the Office of Price Administration during World War II.

Source: NRA Bituminous Coal Statistics, cited in Berquist and Associates, *Economic Survey*, p. 313

The effect of the 1941 changes on the relative costs of the competing districts is more evident. As can be seen from the data in Table 11, average labor costs in Districts 7 and 8 were increased in greater amount than elsewhere, in both relative and absolute terms.

Union wage policy after 1933 caused a significant change in competitive cost relationships between firms in the northern and southern regions of the Appalachian area. In part, the deterioration of the southern position was a result of the elimination of regional wage-rate differentials. Partly, too, the inferior geological conditions in many southern coal fields prevented the use of the most efficient mining machinery; gains in productivity in such mines were smaller than elsewhere. As

3. A comparison of unit costs in May and December is not altogether fair, since the volume of output has some influence on unit costs. The seasonal nature of bituminous mining results in substantially greater activity in the winter months than at other times. There are, unfortunately, no data for any other precode month, nor can the figures for May 1933 be fairly compared with a postcode month.

Table 11. Estimated Changes in Unit Labor Costs, as a Result of 1941 Wage Adjustments

Producing District	*Labor Cost before Change*	*Increase from General Change*	*Increase from Equalization* *	*Total Incr.* †	*Pct. Incr.*
E. Penn. (1)	$1.46	20.7¢	—	22.9¢	15.7%
W. Penn. (2)	1.27	20.2	—	21.9	17.2
No. W. Va. (3)	1.03	16.9	—	17.9	17.4
Ohio (4)	1.02	14.9	—	16.6	16.2
Panhandle (6)	1.04	15.4	—	16.8	16.2
So. No. 1 (7)	1.28	22.3	3.6¢	27.6	21.6
So. No. 2 (8)	1.19	20.4	3.6	25.6	21.6

* North-south differential of 40¢ per day was eliminated.

† Includes cost increase from paid vacations clause.

Source: W. L. Bowden, "Wage and Price Structure of the Bituminous Coal Industry," *Monthly Labor Review, 53* (1941), 308

a consequence, the southern mines were less able to absorb increases in wage rates. In 1934, to summarize the empirical data, southern Districts 7 and 8 were among the three lowest-cost producing areas in the Appalachian region; by 1945 they were among the three highest-cost areas.[4]

The competitive cost position of southern producers was on the

Table 12. Average Labor Cost in Selected Appalachian Districts, 1934 and 1945

Producing District *	*District Average 1934*	*Rank* †	*District Average 1945*	*Rank* †
E. Penn. (1)	$1.22	2	$2.19	2
Md. and Upper Potomac (1)	1.25	1	—	—
W. Penn. (2)	1.14	3	1.92	4
No. W. Va. (3)	.93	7	1.59	5
Ohio (4)	1.09	4	1.48	6
So. No. 1 (7)	1.09	4	2.37	1
So. No. 2 (8)	1.03	6	1.98	3

* Digits in parentheses indicate OPA districts corresponding to titled NRA districts.

† High to low.

Source: National Bureau of Economic Research, *Report of the Committee on Prices in the Bituminous Coal Industry* (New York, 1939), p. 23; Office of Temporary Controls, OPA Economic Data Series No. 2 (Washington, Government Printing Office, 1946), p. 7

4. Figures for 1934, rather than 1933, are used here because there are no data for Southern Numbered 1 in the earlier year.

average, then, worsened by the wage policy of the UMW. Did the changes in the cost structure affect 1. the number of competing mines; 2. the rate and extent of technological change; 3. the number of men employed; and 4. the market position of the northern and southern Appalachian areas?

Effects of Wage Changes on the Number of Mines

As is typically the case in other industries, average total costs of bituminous mines vary over a broad range. There are some low-cost firms, some high-cost firms, and a large number of firms with costs between the two extremes. F. A. Taussig's well-known "bulk-line" cost curve very aptly describes the distribution of coal mines according to average total cost.[5] An indication of the differences in costs among coal mines is given in Table 13.

Table 13. Labor and Total Producing Costs per Ton, 1934

Producing District	*Labor Cost* Average for District	Range for Component Fields	*Total Cost* Average for District	Range for Component Fields
E. Penn. (1)	$1.22	$1.13–1.52	$1.83	$1.65–2.35
Md. and Upper Potomac (1)	1.25	1.25–1.26	1.77	1.72–1.82
W. Penn. (2)	1.14	1.00–1.30	1.74	1.39–2.11
Ohio (4)	1.09	1.05–1.22	1.57	1.48–1.80
No. W. Va. (3)	.93	.89–1.23	1.38	1.27–1.88
So. No. 1 (7)	1.09	1.02–1.20	1.68	1.56–1.83
So. No. 2 (8)	1.03	.78–1.21	1.55	1.23–1.83

Source: NRA Bituminous Coal Code Statistics, cited in National Bureau of Economic Research, *Report of the Committee on Prices in the Bituminous Coal Industry*, p. 23

Given a cost array of this kind, any increase in total costs will have serious implications for marginal operators. Product demand remaining the same, a rise in total costs per ton may transform marginal into submarginal mines. If, as is frequently the case, unit costs are relatively high because of poor geological conditions, there are few ways for high-cost firms to effect compensating economies. Rising costs will tend to drive them out of active operation or at best to reduce their profit margins.

A large proportion of mines in the south are located in areas of comparatively poor geological conditions. These mines have formed the

5. F. A. Taussig, "Price-Fixing as Seen by the Price Fixer," *Quarterly Journal of Economics*, 33 (February 1919), p. 219.

hard core of resistance to union wage demands—particularly since the union achieved the power essential to eliminating the north-south wage rate differential. The high-cost producers have been supported by other southern firms, better endowed geologically, who have profited heavily in the past from the lower wage scale which prevailed generally in the southern Appalachians. Northern producers, both low- and high-cost, have, as a result, had a common interest with the UMW in the elimination of regional differentials in wage rates.

Have rising wage rates—and the elimination of the regional differentials—eliminated any of the existing firms or discouraged entry of new firms? The available data permit no firm conclusions on either question. As can be seen in Table 14, the number of active mines increased steadily during 1935–48. This was, of course, a direct outgrowth of the high and rising level of product demand throughout the period.

Table 14. Number of Active Mines, Selected States, 1935–48

Year	*Penn.*	*Ohio*	*W. Va.*	*Va.*	*Tenn.*	*Ky.*
1935	1365	734	746	82	104	495
1937	1337	736	754	127	130	495
1940	1414	624	743	128	124	631
1942	1742	678	896	152	174	898
1944	1864	553	1031	138	151	1287
1946	1994	567	1126	156	113	1637
1948	2368	697	1423	247	156	2516

Source: U. S. Bureau of Mines

Though there were variations from year to year in the number of active mines in all districts, there is no observable relationship between the worsening of the cost position of the southern mines, relative to the north, and the mine population. Only the state of Ohio in 1948 had fewer active mines than in 1935. On the other hand, the southern states of West Virginia and Kentucky experienced increases of 677 and 2,021 mines respectively.[6]

These data cannot be regarded as conclusive. Without doubt mines

6. Data by states, rather than by districts, have been used because the Bureau of Mines has more consistently reported mine population by states. For purposes of comparison, mine population changed as follows between 1938 and 1948: Districts 1 and 2, increase from 1,295 to 2,368; District 3, increase from 175 to 495; District 4, increase from 599 to 697; District 7, increase from 255 to 437; and District 8, increase from 704 to 2,964. Data from U. S. Bureau of Mines, *Minerals Yearbook* (Washington, 1940), p. 781; and *Mineral Industry Surveys,* Mineral Market Survey 1807 (November 4, 1949), Table 20, p. 28.

were withdrawing from active operation throughout the whole period. The table suggests, however, that other operators were entering the industry at a faster rate than the dying mines were withdrawing. Moreover, it cannot be concluded that union policy was the only or even the major cause for the elimination of those mines which did withdraw. Shortages of labor in time of war, exhaustion of coal deposits in certain fields, inefficiency of mine management—any or all of these could have caused firms to shut down. In any event, the data indicate that high levels of product demand made possible a net increase in the business population. At the same time the possibility must be admitted that the number of new entrants was less than it would have been had union power created fewer uncertainties for the mine operators. On this latter score there is no evidence to support any firm conclusion.

Effect of Wage-rate Changes on Technological Improvement

Though producers of coal-mining machinery had marketed mechanical loading devices as early as 1921,[7] devices of this sort did not enter general use until 1935. The reluctance of coal operators to adopt loading machines did not represent a prevailing bias against modern methods—though doubtless some operators were so biased. Machinery was widely in use in 1921—nearly two-thirds of the national output was being cut by machine. Two considerations, more basic than traditionalism or inertia, accounted for the delay in general installations of loading machinery.

First, the period 1921–35 was one in which the industry as a whole suffered heavy financial losses. Funds were simply unavailable for capital outlays on a large scale: retained earnings of coal firms were small at best, negative at worst; the prospects of future earnings were so poor that neither institutional lenders nor private investors could be lured into making funds available.

Second, the largest single item of producing cost, wage payments, was declining steadily as a result of interregional sales competition. For most coal producers the problem was not reducing costs of production but an expanding demand for their product. The chief exceptions to this generalization were operators of underground mines who were competing primarily with surface mines—a situation confined largely to Illinois and Indiana.

Beginning in 1935 there was a great increase in the use of mechanical loading devices throughout the Appalachian coal region. Particu-

7. The Joy Manufacturing Co. offered for sale a loader which swept coal onto a conveyor chain by means of two steel gathering arms that thrashed outward and back. "Continuous Coal Mining," *Fortune* (June 1950), p. 118.

Table 15. Growth in Tonnage Loaded by Machine, by States, 1926–35

State	Millions of Tons 1926	1928	1932	1935	Per Cent Increase 1926–35
Illinois	2.0	3.2	7.7	12.8	540
Indiana	2.2	2.2	2.4	5.0	127
Pennsylvania	0.8	1.5	1.8	1.4	75
Kentucky	0.2	0.5	0.1	—	—
West Virginia	2.0	2.0	0.6 *	1.2	−40
Alabama	1.2 *	0.8 *	0.2	0.3	—

* Includes the state of Virginia.

Note: The data exclude tonnage handled by conveyors and pit-car loaders.

Source: U. S. Bureau of Mines, as reported in UMW *Journal,* June 15, 1929, p. 12 for 1926–28 figures; *Minerals Yearbook* (1933), p. 406, for 1932 figures; and National Coal Association, *Bituminous Coal Data, 1935–1948,* p. 39, for 1935 data

larly large gains in the percentage of output loaded by machine were registered in the southern fields.

Table 16. Growth in Tonnage Mechanically Loaded, by States, 1935–47

State	1935	1938	1941	1944	1947	Per Cent Increase 1935–47
Pennsylvania	6.5	12.1	35.2	48.5	55.4	816
Ohio	1.5	3.9	11.9	15.1	14.5	867
West Virginia	2.1	20.5	57.5	80.0	99.8	4900
Kentucky	0.5	3.3	11.5	23.4	30.3	3000
Tennessee	0.2	0.5	1.4	2.5	2.8	1300

Note: Includes loaded by machines and handled by conveyors.

Source: U. S. Bureau of Mines, as reported in National Coal Association, *Bituminous Coal Data, 1935–1948,* p. 38

The very large percentage changes noted in Kentucky and Tennessee reflect, admittedly, the very low levels of tonnage loaded by machines in those states in 1935. Even so, as the data in Table 17 reveal, the rate of increase in both southern states in the period 1935–47 was greater than in Pennsylvania and Ohio.

A number of factors contributed to the expansion in mine mechanization. They include: changes in the level and structure of wage rates; rising product demand and a consequent improvement in the earnings prospects of the industry; competitive pressures upon the operators, originating from both within and without the industry; and availability of newer types of equipment, adaptable to a greater variety of geological conditions.

Table 17. Percentages of Output Mechanically Loaded, by States, 1935 and 1947

State	*Total Mech. Loaded* 1935	1947	*Total Output* 1935	1947	*Per Cent Mech. Loaded* 1935	1947
Pennsylvania	6.5	55.4	91.4	147.1	7.1	37.5
Ohio	1.5	14.5	21.1	37.5	7.1	38.6
West Virginia	2.1	99.8	99.2	176.1	2.1	56.8
Kentucky	0.5	30.3	40.7	84.2	1.2	36.0
Tennessee	0.2	2.8	4.1	6.2	4.9	45.0

Note: All figures are in millions of tons. They include tonnage loaded by machines and handled by conveyors.

Source: ibid., pp. 9, 38

Without question the major factor underlying the heightened pace of mine mechanization was union wage policy. Four analysts in the U. S. Bureau of Mines reported in 1938:

It is well-known that the proportion of underground output obtained by mechanical loading has been highest in the coal fields of the northern Rocky Mountains and the Middle West, where high [wage] rates combined with favorable seam conditions have stimulated the process of mechanization. In the last two years, however, market conditions and the trend of wage rates have tended to stimulate mechanization in the Appalachian region, and a large part of the sales of equipment reported by manufacturers went to the Eastern and Southern fields.[8]

With regard to the rapid increase in mechanical loading in the southern Appalachian districts, state officials in West Virginia pointed out in 1939:

Although the Fairmont field of northern West Virginia was one of the earliest regions to experiment on a sizable scale with the then new-fangled idea of loading coal with machinery, and purchased a number of loading units between about 1920 and 1924 . . . it was not until 1936 that it went over permanently to the mechanized side. Mechanical difficulties and relatively low wage scales resulted in the rapid elimination of all the earlier loading machines and conveyors in the district, beginning about 1925, and when these were gone, no new loaders appeared until 1935.

8. L. N. Plein et al., "Loading Machine Sales in 1937," *Coal Age, 43* (1938), 57–8.

While other reasons, such as improvements in equipment, etc., played a part in the rebirth of mechanical loading in West Virginia, *the major factor was rising wage rates,* which forced the adoption of machinery to prevent a rise in mine prices and the consequent loss of business. [Italics added.] [9]

No mine can be mechanized, no matter what the pressures of producing cost, if the available equipment cannot be adapted to the physical structure of the mine. The quickened pace of mechanization in the southern Appalachian region cannot, then, be explained solely as the result of rising wage rates and labor cost per ton. "During 1935 the mobile loader moved into territory hitherto closed to it because of height limitations. Reductions in the vertical dimensions of new models offered during the year opened up new possibilities for mechanization of thinner seams, which was immediately reflected in the installation of machines in such coal by a number of companies throughout the country." [10]

In 1938 *Coal Age* reported: "The height of coal in which mobile loaders can be introduced was subject to still further reduction. . . . As an example, the Stith Coal Company, America, Alabama, began the use of Joy, Jr. loaders with chain conveyors . . . to recover coal averaging around 30 inches in thickness." [11]

Auxiliary equipment was undergoing rapid change at the same time. Locomotives for mine haulage were designed for operation in seams as low as 42 inches. Heavier and better tracks were being built to withstand the heavier loads imposed by mechanical loading. Mines which were ill-adapted to these devices could engage in "trackless mining"—using battery-powered tractors to pull bottom-dumping trail cars, all running along the mine bottom on rubber tires.[12]

In large part, the high rate of capital investment after 1935 was the result of competitive pressures within the industry. The innovators —a few leading firms—naturally improved their competitive positions. In order to restore the previous balance of power their competitors were impelled to follow suit. This is strongly implied in these comments by a leading coal operator, addressing a Congressional committee in 1940: "I have no hesitation . . . in asserting the broad and fundamental proposition that mechanization in the coal industry to date has been a matter of necessity rather than a matter of choice. It has come about not for the purpose of reducing employment, nor for the purpose of increasing the producers' margin of profit. It has

9. "Mechanical Mining in West Virginia," *Coal Age, 44* (1939), 51–3.
10. *Coal Age, 41* (February 1936), 56.
11. Ibid., *43* (February 1938), 53.
12. Ibid., pp. 50–9.

been in order to enable the producer to remain in business and to keep his mines open." [13]

The principal forces underlying the boom in mechanical loading after 1935, then, are two: rising wage rates and labor cost on the one hand, and availability of new equipment adaptable for use in the thin seams on the other. In addition, there were market forces and internal competitive pressures at work which interacted with the two principal factors.

Rising wage rates and labor cost per ton for all firms exerted upward pressures on product prices, intensifying coal's struggle with competing fuels. For all firms this created a strong motive for adopting cost-reducing innovations. In addition, the drastic revision of wage-rate differentials which took place during and after 1933 changed long-standing cost relationships among firms in the northern and southern Appalachian coal fields. This adjustment created a potent incentive to mechanize, especially in the south, and as the innovators made their moves to install new devices, they set up competitive pressures which compelled all firms who would survive to imitate them.

Effects of Wage-rate Changes on the Mine Labor Force

It is a familiar generalization that rising wage rates will stimulate the introduction of "labor-saving machinery." A number of important assumptions are implied: demand for the product is unchanged; production functions of the firms are the same; other factor prices are constant; the elasticity of substitution of productive services is positive, and so on.

Some or all of these conditions may prevail in the short run. It is much less likely that any of them will obtain in the long run. In the present context the chief concern will be with the effects of union policy on employment over long periods, since the strategy of the UMW is to effect a *permanent* reduction in the mine labor force. These kinds of questions must be raised: Have increasing wage rates accelerated the secular decline in mine employment? If so, have there been differences in the employment effects as among producing fields? What effects have rising wage rates had upon the rate of entry of labor into the bituminous industry? What is the significance of the miners' health and welfare fund? What long-run adjustments, in sum, are taking place in the size and composition of the mine labor force and to what extent are they chargeable to union wage policy?

13. C. O'Neill at Hearings, Temporary National Economic Committee, *Investigation of Concentration of Economic Power,* 76th Congress, 3d Session (1940), Pt. 30, p. 17516.

WAGE RATES AND MINE EMPLOYMENT

In a previous chapter it was demonstrated that wage rates of bituminous miners rose more rapidly than elsewhere in the economy during 1933–46. The data in Table 18 indicate, however, that the total number of men employed in the bituminous industry rose only slightly between 1935 and 1948, while in other sectors the increase in total employment was substantial.[14]

Table 18. Average Number of Full- and Part-time Employees in Selected Industries, 1935–48

Year	*Bituminous Coal*	*All Manufacturing*	*Iron and Steel*	*Automobiles*
1935	443	8,904	996	464
1936	457	9,645	1,147	492
1937	470	10,591	1,317	580
1938	416	9,131	1,026	363
1939	381	9,967	1,155	467
1940	439	10,882	1,331	543
1941	452	13,137	1,641	655
1942	480	15,284	1,960	575
1943	434	17,402	2,460	325
1944	415	17,050	2,424	341
1945	388	15,186	2,072	308
1946	391	14,493	1,670	667
1947	429	15,215	1,863	749
1948	452	15,285	1,872	766
Increase 1935–48:				
Absolute	9	6,381	876	302
Relative	2.3%	72.0%	87.9%	65.1%

Note: All figures are in thousands of men.

Source: U. S. Department of Commerce, *Survey of Current Business, National Income Supplement, 1951,* pp. 182–3

Although output of bituminous coal expanded by 60% between 1935 and 1948, three factors made sizable additions to the work force unnecessary. First, there was a marked increase in output per man per day during the period. Second, the industry, at least until 1941, had a large amount of unused or partially used capacity upon which

14. Since part-time operation is much more characteristic of bituminous coal mining than of manufacturing industry, the inclusion of part-time employees in Table 18 inflates the bituminous coal figures more than those of other industries.

to draw. More intensive use of the existing labor force thus permitted substantial increments to output. Third, new mines entering the industry were able by extensive use of machinery to function with far fewer men than were required by older, hand-loading mines. This was particularly true of new surface mines, which entered the industry in great numbers after World War II.[15]

The underlying importance of technological changes—and the demonstrated relationship between rising wage rates and mine mechanization—tempt a conclusion that union wage bargains at least indirectly prevented the expected expansion in the mine labor force. Changes in the number of men employed in the northern and southern regions of the Appalachian fields, however, preclude such a finding. Though wage rates, labor cost per ton and the rate of mechanization all rose more rapidly in the south than in the north, the figures in Table 19 reveal that employment generally increased in the south during 1936–48 and decreased in the north.

Table 19. Average Number of Men Employed, by Districts, 1936–48

Producing District	*Thousands of Men*		*Change, 1936–48*	
	1936	*1948*	*Absolute*	*Per Cent*
1. Eastern Pa.	57.4	51.5	−5.9	−10.2
2. Western Pa.	74.0	57.1	−16.9	−23.0
3. Northern W. Va.	21.5	28.7	7.2	33.5
4. Ohio	29.8	21.1	−8.7	−29.2
6. Panhandle (W. Va.)	4.6	3.5	−1.1	−24.0
7. Southern No. 1	53.0	51.0	−2.0	−3.7
8. Southern No. 2	95.1	121.8	26.7	28.0

Source: U. S. Bureau of Mines, from compilations in National Coal Association, *Bituminous Coal Annual, 1949,* p. 20, and *Bituminous Coal Data, 1951* (Washington, 1952), p. 16

Whatever impact rising wage rates may have had upon the number of men employed, the effects are not identifiable. Doubtless the most important cause for changes in employment in the north and south was variations in product demand.[16]

15. Between 1935 and 1948 the percentage of national output mined by stripping increased from 6.4% to 23.3%. U. S. Bureau of Mines, reported in National Coal Association, *Bituminous Coal Data, 1951.* Table 1, p. 3. The Keystone Coal Buyers' Annual estimated that new strip mines opened between 1946 and 1950 had an annual capacity of 50,000,000 tons. Ibid., note to Table 22, p. 63.

16. This point is explored in a later section of this chapter.

THE AGE COMPOSITION OF THE MINE LABOR FORCE

Basic to the vitality of any industry is a steady inflow of younger men. For this reason it is relevant to inquire into the effects of union wage policy on the age composition of the mine labor force. Were younger men entering the industry at a rate at least equivalent to the rate of separation of the superannuated and disabled? Or did union-management contracts contain provisions which operated to the disadvantage of younger men?

The age composition of the mine labor force is indicated in the data in Table 20.

Table 20. Age Composition of the Mine Labor Force, in Selected States, 1930, 1940, and 1950

	PERCENTAGE BY AGE GROUPS							
State	*17 or Less*	*18 to 24*	*25 to 34*	*35 to 44*	*45 to 54*	*55 to 64*	*65 or Above*	*Total (thous. men)*
Pennsylvania *								
1930	2.9	18.9	23.0	26.6	19.0	7.9	2.0	269.4
1940	0.5	14.2	27.5	23.0	22.4	10.6	2.1	203.0
1950	0.2	8.6	24.7	28.0	21.2	15.4	1.8	173.2
Ohio								
1930	2.6	15.3	19.4	24.5	23.0	12.0	3.3	27.4
1940	0.4	15.7	27.0	20.0	20.9	13.5	2.6	23.0
1950	—	11.5	26.3	27.7	19.3	13.2	2.4	21.4
West Virginia								
1930	2.6	21.1	29.5	25.7	15.0	5.1	1.1	97.3
1940	0.2	16.8	31.6	25.9	17.5	6.9	1.3	105.4
1950	0.2	15.0	30.0	27.6	17.8	8.8	1.4	125.9
Kentucky								
1930	3.4	25.4	31.6	22.5	12.4	4.3	0.9	53.9
1940	0.7	17.9	33.5	26.8	15.0	5.4	1.1	54.3
1950	0.9	16.6	30.5	27.5	17.2	6.6	1.7	63.6
Tennessee								
1930	4.6	25.3	28.7	19.6	15.0	5.8	1.2	8.7
1940	1.6	21.3	32.0	23.4	14.9	6.4	1.6	9.4
1950	1.9	15.1	28.4	30.5	17.4	6.5	1.9	9.2

* Includes employment in anthracite mines.

Source: U. S. Bureau of the Census, *15th Census of the US, 1930,* Population, *4,* Occupations by States, Table 11, pp. 598, 1272, 1406, 1526, 1741; *16th Census of the US, 1940,* Population, Third Series, "The Labor Force," Table 19 (Tenn., p. 87; Penn., p. 97; W. Va., p. 39; Ky., p. 53; Ohio, p. 140); and *17th Census of the US, 1950,* preliminary summaries provided me by Howard G. Brunsman, Chief, Population and Housing Division, December 17, 1952

The age composition of the mine labor force underwent a decided change between 1930 and 1950. In each of the five states considered the percentage of miners of age 24 or less declined. Concurrently, the percentage of workers of age 55 or more increased in each of the states. It may be noted, as well, that the upward shift in the age composition of the work force was absolute as well as relative.

The trend toward an older working group was accelerated during World War II. Military demands on the youth of the nation bit deeply into the bituminous industry. Furthermore, lucrative jobs in other industries beckoned to younger men, particularly those with no deep roots in the coal towns. And rising earnings in the mines made it possible for fathers to send their sons away from the hazardous coal industry to learn other vocations.

With younger men being drawn out of the mine labor force and older men either re-entering the industry or deferring their retirement, the median age of workers rose significantly during the war, although not as rapidly as elsewhere in the economy.

Table 21. Median Age of Male Employees in the United States, 1939–47

Year	*Bituminous Industry*	*All Industries in U.S.*	*Bituminous Age Differential*
1939 *	36.9	32.4	4.5 (years)
1943	40.1	36.9	3.2
1944	41.2	38.6	2.6
1945	41.0	38.3	2.7
1946	39.4	35.7	3.7
1947	38.7	35.6	3.1

* Includes both male and female office workers.

Source: U. S. Social Security Administration, as reported in National Coal Association, *Bituminous Coal Annual, 1950* (Washington, 1950), Table 72, p. 144

At the end of the war many of the younger men returned to the mines and some of the older workers withdrew from the industry. The median age of bituminous miners, of course, began to decline toward the prewar level. Nonetheless, the bituminous work force was in 1947 composed of men significantly older than was common elsewhere in the economy. Less than 14% of bituminous miners were 24 years of age or less in 1947, compared to 22% in all other industries.

While the war had a strong impact on the age composition of the mine labor force, it must be remembered that between 1930 and 1940 the median age of miners was rising. At least two reasons for this trend can be identified. First, the UMW during the thirties sought to "ration" employment by excluding child labor.[17] Laws enacted during the depression years, with the vociferous support of the UMW, had the effect of reducing the number of miners in the Appalachian region aged 17 or less from more than 13,000 in 1930 to scarcely more than 1,000 in 1940.

Table 22. Age of Male Employees in 1947

Years of Age	*Bituminous Industry*	*All Industries in U.S.*
Under 20	2.8%	8.5%
20–24	10.8	13.6
25–29	13.3	13.7
30–34	13.6	12.8
35–39	12.8	11.4
40–44	10.6	9.9
45–49	9.5	8.4
50–54	9.2	7.2
55–59	8.5	6.1
60–64	5.5	4.4
Over 64	3.4	4.0

Note: In 1947 the number of male employees in the bituminous coal industry included in the table above was 570,000, and for all U. S. industries, 32,800,000.

Source: Ibid., Table 73, p. 144

Second, the UMW enforced strictly during the thirties the traditional seniority rules in the industry.[18] The burden of both cyclical and (to a lesser extent) technological unemployment was shifted to the younger men. In the five Appalachian states there was between 1930 and 1940—a period of fluctuating demand for labor—an increase of 4,000 in the number of employed miners aged 55 and above.

The trend in the industry, in sum, has been toward a gradually aging labor force, primarily because the rate of entry of younger men has been less than the rate of attrition of the superannuated and disabled. The union is aware of this; said its president in 1948:

17. A clause in the collective agreement, labeled "Boys," reads: "No person under seventeen (17) years of age shall be employed inside any mine; provided, however, that where a state law provides a higher minimum age, the state law shall govern."

18. A seniority clause was written into the Appalachian contract in 1941. It reads: "Seniority, in principle and practice as it has been recognized in the industry, is not modified or changed by this Agreement. Seniority affecting return to employment of idle employees on a basis of length of service and qualification for the respective positions brought about by different mining methods or installation of mechanical equipment is recognized. Men displaced by new mining methods or installation of mechanical equipment so long as they remain unemployed shall constitute a panel from which new employees shall be selected."

. . . out of 140,000 young men in the armed services who came out of mining communities, it is questionable whether 50 percent of them returned to the mines, because of better opportunities, friends and contacts elsewhere. So our industry is facing acutely the problem of lack of manpower for the future. That's one thing that our country can't permit to happen because our country depends so fundamentally upon the coal industry.[19]

The failure of experienced miners to return to the mines and the apparent unwillingness of inexperienced youths to enter the industry can mean either of two things. It could, on the one hand, be taken as evidence that the union's effort to raise wage rates by reducing employment opportunities in the industry has been successful. Or, on the other hand, it could be a further indication that wage rates had continually to be raised to counteract the attrition on the labor force caused by the plenitude of employment opportunities elsewhere in the economy. As noted in Chapter 6, no firm conclusion on this issue seems justified.

EFFECTS OF THE WELFARE FUND ON THE LABOR FORCE

Apart from the obvious political benefits to the union leaders vis à vis the members, the miners' health and welfare fund is a significant instrument for the advancement of the union's long-run policy objectives. The effects of the fund are to increase the possibility of a reduction over time in the number of mines and miners.

The welfare fund formally began to operate in 1946, under the terms of the "Krug-Lewis agreement."[20] Two aspects of its operation are specially important: 1. the method of financing and 2. the eligibility requirements for pensioners.

The fund is of the noncontributory type, i.e., it is financed exclusively by royalty payments levied on each ton of coal produced. In the initial agreement the royalty was set at 5¢ per ton; by 1948 it had risen to 20¢. This charge is, of course, a direct addition to total producing cost. One of its effects, then, is to weaken further the com-

19. "Pensions, Coming Issue in Labor Relations," an interview with John L. Lewis, *U. S. News and World Report,* November 19, 1948, p. 37.

20. The mines were at the time under seizure by the Federal Government, as a result of an executive order by President Truman invoking the Smith-Connally Act for labor disputes in wartime. Julius A. Krug, then Secretary of the Interior, was entrusted with the operation of the mines. The Krug-Lewis agreement, signed on May 29, 1946, provided in Section 4 for a "health and welfare program." The bituminous coal operators did not recognize the agreement, however, until July 7, 1947. Hearings, Senate Committee on Banking and Currency, *Economic Power of Labor Organizations,* Pt. 1, pp. 410, 427.

petitive position of the high-cost firms—a situation fully in accord with over-all union policy.

The method of financing, moreover, fits in neatly with the UMW's enthusiasm for a smaller labor force.[21] The more general method of financing retirement pension plans is for the employer to pay into a fund a given sum for each employee, whether matched by the worker or not. The larger the work force, the greater the resources of the fund at any given rate at which contributions must be made. The income to the miners' welfare fund is, on the other hand, independent of the number of men employed. So long as productivity increases, the income of the fund will rise—or at least will not fall—even if total employment diminishes.

The eligibility requirements are equally interesting. To qualify for a pension a miner must be a member of the UMW, with a minimum of 20 years' service in the bituminous industry. The union, in addition, has indicated an interest in lowering the retirement age as time passes.[22] These requirements have effects of two different kinds. The provision that the miner must have 20 years of service tends to reduce the mobility of labor, in that the worker is offered an incentive to remain in the industry—movement elsewhere involves the forfeiture of all rights.[23] The union's interest in an earlier retirement age, on the other hand, is completely consistent with its wish for a smaller labor force in the mines. On this second point, however, it must be noted that retirement at age 60 is permissive, not mandatory.

Effect of Wage-rate Changes on Market Position

Survival of firms in the face of rising costs is, in the last analysis, dependent upon the sales of the product. While it is of interest, therefore, to inquire into the effects of changes in wage rates on producing costs, the number of mines, and the volume of employment, all these are closely bound up with the effects of wage-rate increases on the market position of the firms involved. The ultimate question, then, is the extent to which rising wage rates and the elimination of regional wage-rate differentials have influenced interregional competition in the Appalachian area.

21. This point was first suggested to me by N. W. Chamberlain.

22. The retirement age was originally set at 62 years under Resolution 20 of the Trustees of the UMW Welfare and Retirement Fund, dated April 12, 1948, cited in Hearings, Senate Committee on Banking and Currency, *Economic Power of Labor Organizations*, pp. 435–7. It was lowered to 60 by Resolution 8, dated April 7, 1949. Ibid., p. 437.

23. If a worker moves from one mine to another in the industry, he retains all rights to a pension.

Data on total output in northern and southern regions indicate clearly that, contrary to expectation, the southern operators as a group have more than held their own. Total output from the southern states has increased, relative to the north, despite the transformation in the producing-cost structure of the industry. Southern mines, moreover, increased their output despite transportation charges which increased for them more than proportionately to the north after 1945.[24] It must be presumed, in short, that southern mines were able to find markets for an increasing amount of coal, despite their relative disadvantage in production costs and, of course, selling prices.

Table 23. Per Cent of Total Appalachian Output of Northern and Southern Districts, 1935–50

Year	*Northern Districts Nos. 1, 2, 4*	*Southern Districts Nos. 3, 7, 8*
1935	44.7	53.1
1936	44.3	54.6
1937	44.2	54.3
1938	41.1	57.6
1939	41.8	56.9
1940	43.0	55.6
1941	44.0	54.6
1942	43.8	54.9
1943	43.0	55.6
1944	42.8	56.0
1945	43.2	56.0
1946	43.3	56.8
1947	41.7	57.1
1948	41.4	57.8
1949	39.5	59.1
1950	39.6	59.2

Source: U. S. Bureau of Mines, from compilations by National Coal Association, *Bituminous Coal Data, 1935–1948*, p. 10; and *Bituminous Coal Data, 1951*, p. 7

Though the south as a unit improved its relative position during 1935–50, District 7 (southeastern West Virginia and Virginia) suf-

24. R. E. Taggart of the Southern Coal Producers' Association told a Presidential board of inquiry in 1949 that "freight rate increases granted over the past four years have been applied largely on a percentage basis rather than on a cents-per-ton basis." He concluded that "the adverse freight rates against southern coal have become more burdensome." From mimeographed copy of Taggart's remarks, provided by Appalachian Coals.

fered a relative decline in output. As the data in Table 24 indicate, production in District 7 was 17.5% of the Appalachian total in 1935, but only 13.3% in 1950. District 3 (northern West Virginia) and 8 (southwestern West Virginia and eastern Kentucky), on the other hand, improved their share of total Appalachian output—at the expense of northern Districts 1 and 2, as well as southern District 7. These latter changes in market status are best explained by close examination of developments in the several major coal markets.

Table 24. Per Cent of Total Appalachian Output of Various Districts, 1935–50

	DISTRICTS							
Year	*1*	*2*	*3*	*4*	*6*	*7*	*8*	*Total* *
1935	14.4	22.1	7.4	8.2	1.6	17.5	28.2	258
1936	13.0	23.5	7.5	7.8	1.3	17.6	29.5	308
1937	13.2	23.0	7.7	8.0	1.3	17.4	29.2	311
1938	14.0	19.5	8.0	7.6	1.7	18.2	31.4	236
1939	13.2	21.3	8.1	7.3	1.5	18.0	30.8	273
1940	13.2	23.0	8.0	6.8	1.2	17.8	29.8	325
1941	14.2	21.9	8.7	7.9	1.4	16.4	29.5	366
1942	14.3	21.6	9.4	7.9	1.2	15.8	29.7	406
1943	14.5	20.6	10.1	7.9	1.2	15.5	30.0	406
1944	14.4	20.6	11.2	7.8	1.2	14.8	30.0	421
1945	14.4	20.5	11.6	8.3	1.0	14.4	30.0	388
1946	14.6	20.0	10.8	8.7	1.1	14.0	32.0	370
1947	13.9	19.5	11.6	8.3	1.1	13.6	31.9	446
1948	14.2	18.2	11.2	9.0	0.9	14.2	32.4	423
1949	12.8	16.5	11.9	10.2	1.0	13.6	33.6	303
1950	12.8	16.6	11.4	10.2	1.1	13.3	34.5	361

* In millions of tons.

Note: Percentages may not add to 100% because of rounding.

Source: ibid.

In large part, the output of the several Appalachian districts is sold "locally," that is, within the district in which it was mined.[25] A substantial share of the tonnage produced in each district, nonetheless, is sold elsewhere in competition with sellers from one or more of the other coal fields. For the Appalachian region there are at least three distinct markets, other than local, in which interdistrict competition is marked: the tidewater ports at Hampton Roads and Norfolk, Vir-

25. See Table 3 in Chap. 1.

ginia; the lake ports at Ashtabula, Lorain, Cleveland, and Sandusky, Ohio (among others); and the coal markets in the Mississippi Valley. In each of these areas northern and southern coal producers have historically been in competition with one another.

Table 25. Per Cent of Total Shipments to Tidewater Ports, 1932–46

Year	*From Pa. and No. W. Va.*	*From Southern Low Volatile*	*From Southern High Volatile*
1932	38.0	43.5	18.0
1933	36.0	47.5	16.5
1934	37.0	47.5	15.0
1935	37.0	48.0	15.0
1936	36.5	48.5	15.0
1937	35.5	49.0	15.5
1938	38.0	48.0	14.0
1939	37.5	49.0	13.0
1940	37.0	50.0	13.0
1941	40.0	46.5	13.0
1942	50.0	42.5	7.0
1943	49.0	46.0	5.0
1944	51.0	41.0	8.0
1945	51.5	35.5	13.0
1946	53.0	31.5	14.5

Source: U. S. Bureau of Mines, as compiled in National Coal Association, *Bituminous Coal Data, 1935–1948,* Table 60, p. 66. Comparable data for the years 1946–50 are not available

The data in Table 25 depict the trends in tidewater loadings from 1932 to 1946. Two points are specially worthy of comment. The narrowing of the range of regional wage-rate differentials in 1933 apparently had little effect on the sales of mines in the southern low-volatile areas. The relative share of those producers increased slowly, but perceptibly, up 1940. Sellers in the southern high-volatile fields, however, suffered a steady decline in sales, relative to other areas, until 1943.[26]

The second point of note is that after 1940 tidewater loadings of the southern low-volatile fields began a steady decline, relative to the northern producers. In absolute terms shipments from the south dropped by 2,000,000 tons between 1940 and 1946, while shipments

26. The basic data indicate, as well, that despite some fluctuations over the period, sales of the southern high-volatile producers dropped from nearly 5,000,000 tons in 1932 to slightly less than 2,000,000 tons in 1943.

from the north rose by 14,000,000 tons. After 1940 and until 1943 tidewater loadings from the southern high-volatile fields also dropped sharply.

Without doubt a number of variables contributed to the changes just noted. Sales of high-volatile coals were certainly affected by the competition of substitute fuels.[27] Governmental regulation of the distribution of bituminous coal during World War II tended to disrupt the normal flow of trade.[28] The sudden rise in demand for coal, brought on by the war, must itself have caused changes in the pattern of distribution as coal consumers sought to hedge their positions by placing orders in several different districts, where previously they had done business in only one area. These and other factors preclude an unqualified conclusion that the shift after 1941 in tidewater loadings to the

Table 26. Per Cent of Shipments by Westbound Rail to Mississippi Valley, 1932–46

PER CENT OF TOTAL SHIPMENTS FROM:

Year	*Ohio*	*Pa.*	*No. W. Va.*	*So. W. Va. High Vol.*	*So. W. Va. Low Vol.*	*E. Ky., Tenn., and Va.*
1932	13.4	13.4	4.8	23.5	23.5	21.5
1933	19.4	10.6	5.0	22.5	23.4	20.6
1934	16.4	17.6	4.9	20.1	21.7	19.6
1935	15.6	15.6	4.3	22.2	21.8	20.5
1936	13.8	18.3	4.0	20.6	22.3	20.7
1937	14.0	17.7	4.1	20.3	23.0	21.1
1938	13.0	15.6	4.2	22.3	21.6	23.0
1939	12.8	17.1	4.5	21.2	22.2	22.0
1940	11.8	18.2	4.2	20.3	23.4	22.0
1941	11.7	18.0	5.1	19.7	24.7	20.6
1942	11.2	17.4	5.0	20.6	24.0	21.8
1943	11.9	17.2	5.9	21.2	23.0	21.5
1944	13.5	18.0	7.3	20.7	20.5	19.7
1945	15.6	18.0	7.5	20.0	19.7	19.1
1946	16.6	17.4	5.8	21.0	19.9	19.6

Source: ibid., Table 64, p. 70

27. High-volatile coal is used for coking, railroad and bunker fuel, and industrial purposes. In these areas interfuel competition has been particularly keen. See Chap. 8. Rising exports to Europe after 1943 probably account for the upward movement in tidewater loadings of southern high-volatile coal in that year and thereafter to 1946.

28. A typical order of the Solid Fuels Administration for War required coal operators in the Appalachians to give first priority to the Great Lakes region, doing so at the expense of shipments to the northeastern U. S. Press Release SFA 123, October 18, 1943. While orders of this sort were for short periods of time, they must have affected the traditional market relationships which had developed in the industry.

Table 27. Per Cent of Shipments to Lake Erie Ports, 1932–46

	PER CENT OF TOTAL LOADINGS FROM:					
				So. W. Va.		*E. Ky., Tenn.,*
Year	*Ohio*	*Pitts.*	*No. W. Va.*	*High Vol.*	*Low Vol.*	*and Va.*
1932	6.3	31.0	5.2	23.8	16.7	17.1
1933	8.1	27.5	4.0	23.0	20.0	17.4
1934	7.2	30.0	3.6	21.6	19.2	17.8
1935	6.4	28.3	3.4	21.8	20.7	19.5
1936	6.4	24.7	3.5	23.1	22.2	20.0
1937	7.1	26.1	5.1	24.4	18.6	18.8
1938	6.8	22.8	4.0	23.6	21.6	21.1
1939	5.8	22.5	4.9	26.5	21.2	19.4
1940	5.4	24.1	5.0	25.0	21.6	19.0
1941	7.6	22.5	5.8	27.8	17.6	18.7
1942	8.5	18.8	5.7	29.7	18.6	18.8
1943	9.9	17.7	5.9	30.0	18.4	18.4
1944	9.0	19.1	6.7	25.0	19.5	21.0
1945	8.4	18.7	7.0	24.0	19.5	22.2
1946	8.9	17.4	6.5	25.9	19.8	21.8

Note: Totals may not add to 100% because of rounding.

Source: ibid., Table 62, p. 68

favor of the northern producers was centrally induced by the elimination of the regional wage-rate differential.

In the other two markets, the Mississippi Valley and the lake ports, the equalization of wage rates in north and south seems to have had little or no effect on coal sales. On westbound rail shipments to the Mississippi Valley, as the data in Table 26 indicate, the northern fields made relatively small gains after 1932. Through 1932–46, however, there were marked fluctuations in the relative shares of the various districts. And the fluctuations occurred without relation to important adjustments in either the regional wage- or freight-rate structures.

The elimination of regional wage-rate differentials had no apparent ill effect on southern shipments to the lake ports. The data in Table 27 show that, on the contrary, the relative share of southern fields increased between 1932 and 1946. The proportion of shipments from the Pittsburgh fields, on the other hand, fell during the period from 31% to 17%.

With the notable exception, then, of the tidewater market, it appears that southern sales, relative to other districts, were not severely damaged by the equalization of wage rates in the Appalachian region. The matter cannot, however, end here. Some account must be taken of postwar developments, on the reasonable presumption that after 1946 the coal trade tended to revert to the prewar pattern of distribution. As product demand receded from its postwar peaks, it was to be expected that the alterations in the interregional cost structure since 1941 would operate to the disadvantage of the southern producers.

Data on bituminous coal distribution to various markets is much less complete for the postwar years than for the period 1932 to 1946.[29] Other statistics are available, though they are not fully comparable with those presented above. As presented in Table 28, the data indicate that the greatest changes in shipments occurred in southern Districts 7 and 8. Nothing in the table suggests, however, that in markets where northern and southern Appalachian producers compete, gains by one area were made at the expense of the other. Quite the contrary, the losses of District 7 were largely offset by increased sales by producers in District 8.

Table 28. Per Cent of Total Rail Coal Shipments into Various Areas from Appalachian Districts, 1941 and 1951

	FROM PRODUCING DISTRICT:											
Rail Shipments into:	*1*		*2*		*3*		*4*		*7*		*8*	
	1941	*1951*	*1941*	*1951*	*1941*	*1951*	*1941*	*1951*	*1941*	*1951*	*1941*	*1951*
Upper Peninsula of Michigan	0.7	2.3	*	0.2	0.6	2.6	0.7	1.8	59.8	30.7	38.3	62.5
State of Illinois	*	*	*	*	0.3	0.2	*	*	23.3	17.5	17.2	20.7
State of Ohio	0.1	1.3	22.7	16.9	7.3	6.9	18.2	25.9	16.6	14.9	33.4	33.6
State of Indiana	*	*	*	*	0.2	0.4	0.5	0.4	8.9	7.3	37.2	29.8
Lower Peninsula of Michigan	0.1	0.3	0.1	0.7	0.5	2.2	7.0	17.2	27.6	19.1	63.2	54.6

* Denotes percentage of less than one-tenth of 1%. So little tonnage was reported as sold from District 6 (Panhandle of West Virginia) that it was not included in the table.

Source: Roy Carson's Report, as compiled by V. M. Johnston, Controller, Appalachian Coals, Inc.

The available evidence indicates that the elimination of the regional differential in wage rates in 1941 and the steady increase in wage rates

29. The year 1946 was the last in which coal operators were required to submit distribution data to the Solid Fuels Administration for War.

since 1933 has had relatively little effect on interregional competition in the sale of the product. Southern producers as a group have, in fact, increased their share of the bituminous coal trade, relative to northern operators. There is but one exception to this general observation: the southern producers since 1932 have held a diminishing share of total shipments to tidewater.

The most important aspect of these developments is that union policy has succeeded in slowing down the rate of expansion of the southern relative to the northern coal fields. The imposition on the industry of a uniform structure of wage rates has resulted to a substantial degree in the success of the union's effort to stabilize interregional competition.

SUMMARY

Mindful of the disastrous interregional competition of the 1920's, the UMW has sought to stabilize competitive relationships between northern and southern Appalachian producers. This it has attempted to do by eliminating regional differentials in wage rates and by exerting steady pressure to raise the level of wage rates. In this way, the union hopes, producing costs for all firms will be raised and the less efficient firms gradually pushed out of the industry.

In this chapter it has been shown that union policy has been a major factor in the transformation of southern mines from low-cost to high-cost operations. Yet during the same period the number of operating mines increased in the south by more than the average amount. In like fashion the upward thrust of producing costs stimulated a general expansion in mine mechanization—with the most rapid rate of technological change occurring in the south. Again, while over-all employment in the bituminous industry rose only slightly between 1935 and 1948, large gains were registered in southern fields as against substantial declines in employment in the northern fields. Finally, the relative share of total Applachian sales by southern firms increased during 1935–48. In only one market, the Atlantic coast ports, did sales of southern bituminous, relative to that of northern producers, decline.

Two clear conclusions seem justified. First, the competitive advantages of southern over northern coal producers was (and is) based on factors other than the wage bill. Even with virtually equal wage rates for both regions, southern producers have improved their relative position, though at a rate much less pronounced than in the twenties. Second, the slower rate at which adjustments in the relative position of the two regions have taken place is in large part a direct result of aggressive union wage policy.

CHAPTER 8

Union Policy and Interfuel Competition

INTRODUCTION

UNION WAGE POLICY is of critical importance only if it affects the ability of firms in the industry to sell what they produce. There is no question but that the wage policies of the UMW have raised average production costs and selling prices of bituminous coal. It remains to be determined whether or to what extent union policy has affected total consumption of coal.

The coal producers have taken the position generally that sales of their product are a function principally of selling price; that is, that the coefficient of elasticity of demand for bituminous is greater than zero. They have insisted as a result that if wage rates are to rise, the increases must be at a rate equal to or, preferably, less than increases in the productivity of labor.

The UMW argues, on the other hand, that a consumer's choice of any fuel is rarely a function of price, but rather is dictated by the user's income and his "taste." In this view a fuel buyer may select a higher-priced source of energy simply because that fuel bestows certain "nonprice" benefits. Moreover, the union insists, for most consumers the price of fuel is but a small percentage of total costs. Changes in fuel prices will, consequently, have a less than proportionate impact on the user's total producing cost. Finally, the union points out, it is the delivered price of fuel which is significant to the prospective buyer. Since transportation charges are often as much as 50% of the delivered price, changes in the wage level will be only partially reflected in market price, as distinguished from mine price (f.o.b.).

The dispute is obviously of vital importance to the industry. The volume of product sales intimately affects the level of profits and the demand for labor. It is the purpose of this chapter to determine, if possible, which, if either, of the conflicting viewpoints is valid. Questions such as these must be considered:

1. What have been the trends in coal consumption in the past thirty years? Is the use of bituminous as an energy source increasing, decreasing, or remaining constant?

2. What have been the trends in use of bituminous coal among its largest consumers? What factors influence the consumers' ultimate decision to burn bituminous coal or some substitute fuel?

3. To what extent is union wage policy responsible for changes in consumption of bituminous coal, relative to other fuels?

Secular Trends in Coal Consumption

In the last three decades consumption of bituminous, relative to alternative sources of energy, has declined. In absolute terms, however, total sales of bituminous have remained at about the same level.[1]

Table 29. Per Cent of Total Energy Supply Provided by Various Fuels in B.T.U., 1920–50

Fuel	*1920*	*1925*	*1930*	*1935*	*1940*	*1945*	*1950*
Bituminous	71.8	70.1	64.0	59.9	59.2	56.1	44.8
Anthracite	11.7	8.7	9.3	8.2	6.4	5.0	3.8
Petroleum	7.0	11.1	13.0	15.5	17.2	19.3	22.8
Natural Gas	4.3	5.9	9.5	11.1	12.3	13.9	22.5
Water Power	5.2	4.2	4.2	5.3	4.9	5.7	6.1

Source: U. S. Bureau of Mines and Federal Power Commission, as compiled in National Coal Association, *Bituminous Coal Annual, 1951* (Washington, 1951), Table 38, p. 94

Coal consumption reached a peacetime peak of 519,000,000 tons in 1929. Thereafter, until the outbreak of World War II, sales of bituminous diminished, while the demand for competing fuels continued to grow. This relative decrease in the use of bituminous cannot be explained by the general depression of the 1930's; the demand for all fuels is derived and each felt the impact of the reduction in industrial activity. The fact is, instead, that between 1920 and 1950—particularly after 1929—coal's competitors usurped most of the increased demand for fuel which accompanied the growth of the American economy.[2]

Two principal reasons account for the decline in the growth of de-

1. Total sales in 1920 were approximately 508,000,000 tons, while in 1950 they were estimated at 454,000,000 tons. It should be noted that the figures given are not the same as for total production, since all coal produced is not sold on the market. See note to Table 29.

2. A substantial part of the increase in demand for energy was in uses where coal could not compete. They include: gasoline and diesel fuel used by trucks, tractors, and automobiles; road oil; lubricants; and natural gas used for carbon black. These markets excluded, bituminous coal supplied nearly 69% of all energy in markets where it competed with all fuels in 1926, but only 46% in 1949. D. R. G. Cowan in lecture to Graduate Economics Club, University of Michigan, December, 1950.

mand for bituminous. First, there have been major advances in the efficiency of fuel consumption. Though total output of goods and services in coal-using industries has increased, the need for coal has expanded in a smaller proportion.

Table 30. Indicators of the Effect of Fuel Economy on Consumption of Coal in the United States, per Unit of Performance, 1919–47

POUNDS OF COAL

Year	*Per kwh Electricity*	*Per 1,000 Gross Freight Ton Miles*	*Per Passenger Train Car Miles*	*Per Gross Ton of Pig Iron*
1919	3.2	164	18.1	3,428
1923	2.4	161	18.1	3,323
1927	1.8	131	15.4	3,094
1931	1.6	119	14.5	2,923
1935	1.5	120	15.5	2,838
1939	1.4	112	14.8	2,853
1943	1.3	114	15.0	2,934
1947	1.3	114	15.9	3,086

Source: U. S. Bureau of Mines, as compiled in National Coal Association, *Bituminous Coal Data, 1935–1948,* p. 110

Second, there has been a steady substitution over time of alternative fuels by industries formerly using large amounts of bituminous coal. The principal consumers of bituminous are: railroads, electric utilities, producers of coke (high temperature and beehive), steel mills, cement mills, and domestic and commercial consumers.[3] Since fuel substitution has been the more dramatic influence in the relative decline in coal demand, it is proposed to examine changes in fuel consumption in all but two of the industries listed. Producers of coke and steel will be omitted, since much of their coal is provided by captive mines and does not, therefore, influence the markets of commercial producers.

SECULAR TRENDS IN FUEL CONSUMPTION BY RAILROADS

There has been a steady reduction in consumption of bituminous by the railroads, particularly since 1936. That date is significant, be-

3. In 1950 electric utilities consumed 88,000,000 tons, or 19% of total tonnage sold in the United States; railroads used 65,000,000 tons, or 14%; coke, gas, and steel producers consumed 111,000,000 tons, or 25%; other industries of all types burned 103,000,000 tons, or 23%; and deliveries to retail dealers amounted to 87,000,000 tons, or 19%. Data from U. S. Bureau of Mines, as compiled in National Coal Association, *Bituminous Coal Annual, 1951,* p. 101.

cause it marks the beginning of the general adoption by American railroads of the diesel-electric locomotive.[4]

Table 31. Per Cent of Various Fuels Consumed by Locomotives, Class I Railroads, 1936–50

Year	*Bituminous Coal*	*Fuel Oil*	*Diesel Fuel*	*Electricity*
1936	81.3	16.4	0.3	2.0
1937	80.4	16.9	0.6	2.1
1938	78.3	17.7	1.2	2.8
1939	78.0	17.2	1.7	3.1
1940	77.5	17.3	2.0	3.2
1941	75.8	17.6	3.6	3.0
1942	74.3	18.9	4.1	2.7
1943	73.4	19.7	4.3	2.6
1944	70.5	19.6	7.2	2.7
1945	67.0	20.0	10.3	2.7
1946	63.8	19.1	14.1	3.0
1947	60.6	17.8	18.7	2.9
1948	53.3	16.0	27.8	2.9
1949	42.9	13.3	40.8	3.0
1950 (Prelim.)	36.6	11.1	49.5	2.8

Source: Interstate Commerce Commission, as compiled in National Coal Association, *Bituminous Coal Annual, 1951,* Table 47, p. 113

The data in Table 31 indicate the rapid increase in the consumption of diesel fuel by Class I railroads, particularly since the end of World War II. The data reflect in part the high volumes of traffic during the postwar period. But as the statistics in Table 32 reveal, it represents, too, a steady displacement of steam (both coal- and oil-burning) locomotives in favor of diesel-electrics.

SECULAR TRENDS IN FUEL CONSUMPTION BY ELECTRIC UTILITIES

Consumption of bituminous coal by elecrtic utilities has increased steadily since 1920. Relative to other fuels, as a result, coal has retained its position in this important market.

4. The diesel-electric engine had been used for locomotive power in Europe as early as 1912. It was introduced into the United States in 1924, being employed primarily as a switching engine. Not until 1933, when the first "streamliner" was put into service, was the diesel an important factor in long hauls. Thereafter, its use expanded rapidly—especially because of the intense competition for traffic among the western railroads. See A. I. Lipetz, *Diesel Engine Potentialities and Possibilities in Rail Transportation,* Engineering Bulletin, Purdue University (1935), pp. 5–8.

Table 32. Kinds of Locomotives in Use by U. S. Railroads, 1924–48

Year	Steam	Electric *	Diesel	Total †
1928	62,600	617	—	63,300
1930	59,400	663	77	60,190
1932	55,800	764	80	56,700
1934	50,500	805	104	51,400
1936	46,900	858	175	48,000
1938	45,200	882	403	46,550
1940	42,400	900	967	44,300
1942	41,700	892	1,978	44,700
1944	41,900	902	3,432	46,300
1946	39,600	867	5,008	45,500
1948	34,600	867	8,981	44,500
PER CENT OF TOTAL LOCOMOTIVES				
1928	98.8	1.0	—	
1930	98.5	1.1	0.1	
1932	98.2	1.4	0.1	
1934	98.0	1.6	0.2	
1936	97.6	1.8	0.4	
1938	97.0	1.9	0.9	
1940	95.5	2.0	2.2	
1942	93.2	2.0	4.4	
1944	90.4	1.9	7.4	
1946	87.0	1.9	11.0	
1948	78.0	2.0	20.0	

* Units.

† Includes other types (such as gasoline).

Source: Interstate Commerce Commission, *Statistics of Railways of the United States* (1941), p. 15, and (1948), p. 14

In the past 30 years relationships among the competing fuels have been virtually unchanged. Only natural gas has made inroads of any size into the utilities market. In 1920 1.5% of total kilowatt hours were produced with natural gas as the energy source; by 1948 natural gas was being used in the production of about 11% of total kilowatt hours.[5] The gains of natural gas were not, however, at the expense of bituminous coal, but of hydroelectric power.[6]

5. Federal Power Commission, as reported in National Coal Association, *Bituminous Coal Annual, 1949,* p. 118.

6. The diminishing importance of hydroelectric power, relative to mineral fuels, has been explained in these terms: ". . . it must be remembered that the [9 million hp.] of water power that has already been developed comprises those sites which could be economically developed. Further expansion of water power plants must be

Table 33. Consumption of Bituminous by Electric Utilities, 1920–48

Year	*Tons of Bituminous Consumed*	*Per Cent of Electricity Produced by Bituminous Coal*
1920	30,100,000	52.8
1925	33,800,000	57.1
1930	38,100,000	55.5
1935	30,900,000	47.6
1940	49,100,000	54.1
1945	71,600,000	51.6
1948	95,700,000	54.1

Source: Federal Power Commission, as compiled in National Coal Association, *Bituminous Coal Annual, 1949,* pp. 112, 118

SECULAR TRENDS IN FUEL CONSUMPTION BY CEMENT MANUFACTURERS

The cement industry has never been a major consumer of bituminous coal, in terms at least of the volumes used by the railroads and electric utilities. Cement manufacturers in the peak year 1927 used approxi-

Table 34. Per Cent of Total Consumption of Various Fuels in Cement Manufacture, 1927–48

Year	*Bituminous Coal*	*Fuel Oil*	*Natural Gas*	*Other*
1927	82.0	8.8	7.5	1.7
1929	79.3	5.2	13.8	1.7
1931	77.5	6.0	14.9	1.6
1933	68.7	9.0	20.7	1.6
1935	71.1	6.7	21.0	1.2
1937	70.3	7.7	21.0	1.0
1939	70.4	7.8	20.8	1.0
1941	69.2	8.6	21.3	0.9
1943	68.5	8.2	23.2	0.1
1945	65.3	11.2	23.3	0.2
1946	68.8	9.3	21.7	0.2
1947	69.2	9.6	21.1	0.1
1948	68.6	8.8	22.5	0.1

Source: U. S. Bureau of Mines, as compiled in National Coal Association, *Bituminous Coal Annual, 1949,* p. 128

made at localities not so suitable for development; hence the initial investment for future developments will probably increase as it becomes necessary to develop the less desirable water powers. On the other hand, the economy of steam power plants is increasing steadily." J. G. Tarboux, *Electric Power Equipment* (3d. ed. New York, McGraw-Hill, 1946) pp. 15–16.

mately 10,000,000 tons. Consumption in 1948 was about 1,500,000 tons less than that top figure. Over the 20-year period, however, there was increasing utilization of natural gas, which was the major cause for a decline in the relative importance of bituminous in this market.

SECULAR TRENDS IN FUEL CONSUMPTION BY DOMESTIC AND COMMERCIAL USERS

Competition among fuels is unusually intense in domestic and commercial uses. Nine mineral fuels compete in this market: bituminous and anthracite coal, briquettes and packaged fuels, manufactured gas, fuel oil, natural gas, liquefied petroleum gases, and petroleum coke. Of these only four are, however, close competitors: bituminous, anthracite, fuel oil, and natural gas.

Table 35. Per Cent of Consumption of Various Fuels for Residential and Commercial Purposes, 1935–49

Year	*Bituminous*	*Anthracite*	*Fuel Oil*	*Natural Gas*
1935	41.9	20.4	11.0	8.3
1936	39.6	19.7	13.4	8.6
1937	37.3	18.4	15.6	9.1
1938	34.3	17.9	17.2	9.7
1939	33.3	18.2	18.1	9.5
1940	36.1	15.7	19.0	9.5
1941	37.6	15.5	18.5	9.1
1942	38.1	15.6	17.5	10.0
1943	41.9	14.6	15.3	10.1
1944	41.1	14.5	14.8	10.3
1945	40.1	12.4	16.0	11.1
1946	33.6	13.0	18.8	12.1
1947	30.7	10.5	21.4	13.5
1948	26.6	10.6	22.8	14.5
1949	27.1	8.1	22.2	16.1

Source: Based on data of U. S. Bureau of Mines and American Gas Association, as compiled in National Coal Association, *Bituminous Coal Annual, 1951,* pp. 134–5

The data in Table 35 indicate clearly the secular decline in use of bituminous coal, relative to fuel oil and natural gas. It may be noted in passing that the brief rise in the percentage of bituminous consumed during 1940–43 reflects primarily the wartime demands for fuel oil in military uses and the restrictions imposed on new construction of natural-gas pipelines.

Factors in the Relative Decline in Demand for Bituminous Coal

Five major influences can be identified as having contributed to the decline in the importance of bituminous coal as a source of energy. They are closely related to one another. Without attempting to list them according to their significance, they are: 1. the greater convenience of substitutes; 2. availability of alternative fuels; 3. more rapid technological improvements in equipment for burning other fuels; 4. self-protection of large consumers against interruptions in their supply of fuel; and 5. cost economies arising from the use of substitute products.

CONVENIENCE OF SUBSTITUTE FUELS

Compared to other sources of energy, bituminous coal is for many uses most inconvenient. Considerable storage space must be provided in the home, the apartment building, the commercial establishment, or the environs of the electric utility or industrial plant. The fuel-oil tank can, on the other hand, be buried underground, while virtually no storage facilities are required for natural gas.

There are, in addition, ashes to be disposed of after coal has been burned—a problem which arises not at all with substitute fuels. Coal is a "dirtier" fuel than either of its competitors, an important factor in homes and business establishments. And bituminous is at a special disadvantage in urban areas where smoke is becoming a problem of increasing concern to the community. Consumers of coal have been under pressures of several kinds to adopt other fuels at the pain of installing elaborate equipment to abate smoke.[7]

Concern for convenience, while important in the decision-making process, is probably not a conclusive reason for the shift to other fuels by railroads and utilities. It is very likely the principal reason for the increasing adoption of substitute fuels in domestic and commercial uses. A study by the Federal Power Commission in 1948 concluded that "there is increasingly evident a pronounced trend toward refined fuels which offer to consumers desired advantages of utilization, this preference being particularly strong with respect to gas fuel for both domestic and industrial purposes." [8]

7. As early as the 1920's railroads were introducing gasoline- and oil-burning locomotives into their freight yards for switching service. This was done to aid in smoke-abatement programs and, as well, because coal-burning locomotives could not be used safely inside warehouses. The extent of the smoke nuisance is illustrated by the conservative estimate that smoke in New York City annually damages personal property at a rate of $100,000,000. E. Hodgin, "Coal. The Fuel Revolution," *Fortune*, *35* (1947), 245.

8. Federal Power Commission, *Natural Gas Investigation*, Docket G-580 (Washington, 1948), p. 339.

AVAILABILITY OF ALTERNATIVE FUELS

The Federal Power Commission has pointed out that "conditions as to interfuel competition depend fundamentally upon the availability of the fuels in particular consuming markets. . . ."[9] It is this consideration which explains in large measure, for example, the rapid expansion in the utilization of natural gas by electric utilities. Despite its inefficiency, compared to other fuels, natural gas provided energy for electric power plants producing 11% of total power output in the United States in 1948.[10] Fuel oil was used in the production of but 6%. The figures are somewhat misleading, however, since most of the natural gas was used in utility plants located in five western states.[11] Each of those states was a major producer of gas, while deposits of bituminous coal there are scanty and/or inferior.

The available data leave little doubt that availability at a price is the central factor in the choice of a fuel. In ten states—Virginia, Wisconsin, Ohio, North Dakota, Indiana, Illinois, Kentucky, Michigan, West Virginia, and Pennsylvania—65% or more of the fuel requirements in 1949 were met with bituminous coal. Each state is, of course, a major producer of bituminous coal or is near a primary source. Eight states—Arkansas, Oklahoma, Oregon, New Mexico, Arizona, California, Louisiana, and Texas—relied on bituminous for 8% or less of their fuel needs. In these states as much as 93% and not less than 54% of fuel requirements were met with natural gas.[12] All eight states have negligible deposits of high-grade coal, and transportation costs make its importation impractical.

TECHNOLOGICAL CHANGES IN FUEL-BURNING EQUIPMENT

Technological changes have played a key role in the relative decline in coal demand. In the 1920's, for instance, the decision in the steel industry to use greater amounts of scrap rather than producing pig iron caused a substantial reduction in the industry's consumption of

9. Ibid., p. 339.

10. A study by the FPC revealed that electric power plants fueled by natural gas consumed nearly 8% more b.t.u.'s than was experienced in plants burning bituminous coal. Thermal efficiency in coal-burning plants averaged 24%; in oil-burning plants, 23.4%; and in gas-burning plants, 22.2%. National Coal Association, *Bituminous Coal Annual, 1949*, p. 118.

11. Nearly 70% of the natural gas consumed by all electric utilities in 1946 was burned in California, Kansas, Louisiana, Oklahoma, and Texas.

12. In Oregon 53% of energy requirements were served by hydroelectric power, and an additional 41% by fuel oil. Data from U. S. Bureau of Mines, Federal Power Commission, and Edison Electric Institute, as compiled in National Coal Association, *Bituminous Coal Annual, 1951*, Table 39, pp. 96–7.

coal. Similarly, boiler steels have steadily been strengthened to withstand increasing steam pressures, thereby contributing to greater efficiency in coal use.

Technical improvements of a different order have furthered the reduction in coal consumption. Notable among these is the diesel-electric locomotive, which is steadily displacing the coal-burning steam engine on the railroads. The diesel is superior to steam locomotives in several respects. First, it is more reliable. An Interstate Commerce Commission study of motive power indicated that diesels averaged 220% more locomotive miles per failure than steam engines.[13] Second, diesels are more continuously available. They demand less maintenance and repair than steam locomotives, and when service is necessary it can be accomplished much more quickly. This means not only that costs of operation tend to be lower, but also that diesels can be operated a greater number of hours per month.[14] Finally, the initial cost of a diesel is now less than that of steam locomotives, each of which is "custom-made." Standardization of design and the application of mass-production techniques have contributed importantly to a reduction in the costs of producing diesels, relative to steam equipment.[15] All these advantages of the diesel have made it ever more attractive to railroad managers—the principal reason for the displacement of bituminous coal in the railroad industry.

In like fashion—though without such striking results—technical changes have advanced the use of competing fuels in uses where coal had once been supreme. In areas where fuel prices are high, large fuel consumers—notably electric utilities—are installing burning equipment which can be adapted on 24-hour notice to any of the three major fuels. In domestic markets technology has overcome the unwillingness of householders to scrap a functioning coal furnace; the furnace can, at comparatively small expense, be converted into an oil burner.

SELF-PROTECTION AGAINST UNCERTAINTIES OF FUEL SUPPLY

So dependent is the economy upon a continuous supply of fuel that even short interruptions in fuel output can cause dislocations. To the domestic consumer a fuel shortage can cause much discomfort; to an electric utility it can mean a reduction or even a cessation of operations. To an extent, consumers of bituminous can hedge against

13. ICC, Bureau of Transport Economics and Statistics, *Study of Railroad Motive Power,* Statement No. 5025 (Washington, May 1950), Table 38, p. 103 (mimeographed).
14. Ibid., pp. 64–5.
15. Ibid., p. 97.

interruptions in supply by maintaining stockpiles. There are, however, limits to stockpiling, set by storage space, willingness or ability to tie up spendable funds, and the expenses of handling and rehandling the stored fuel. This means, then, that consumers of bituminous are vulnerable to work stoppages, whether called at the union's instance or resulting from other causes.

Of all the competing fuels bituminous coal has the worst record in recent years for continuous production. While there is no clear evidence that any consumer of bituminous has ever been forced to close his plant during a strike for complete lack of coal, it is true, nevertheless, that the frequent strikes have created uncertainty and have been a nuisance to coal consumers. It cannot be doubted that many householders have shifted to substitute fuels, even at a higher price, to assure themselves of a reliable and steady supply of fuel.

It is certain, too, that vagaries in supply have encouraged large industrial consumers, especially electric utilities, to diversify their fuel consumption. Consolidated Edison of New York City reported in 1947 that it "was now in a position to carry 25 percent of its normal power load on oil in an emergency and that arrangements for further emergency conversion were being investigated. During the May [1946] coal strike the Commonwealth Edison Company in Chicago made emergency installations of oil-burning equipment to fire seven high-pressure steam generating units. An even more extensive conversion program was carried out during the November [1946] mine strike by the Cleveland Electric Illuminating Company." [16]

In each of the cases cited economy of operation was a major factor in the decision to diversify. But as an official of the Peabody Engineering Company put it, the uncertainty caused by the frequent mine strikes was "the straw that broke the camel's back." [17]

COST ECONOMIES OF SUBSTITUTE FUELS

Although outlays for fuel are typically a small proportion of total producing costs, most industrial consumers are sensitive to changes in relative prices of competing fuels. That is specially true of electric utilities, the selling prices of which are regulated and, therefore, "sticky." During periods of inflation, when prices of productive services are rising more rapidly than regulating agencies are allowing increases in product prices, cost economies must be realized wherever possible. In all industries, moreover, fuel expense is one of the few remaining items of cost subject to day-to-day adjustment.

Relative prices of fuel are rarely the *sole* reason for an option for

16. *New York Times,* January 27, 1947, p. 13.
17. Ibid., p. 13.

one or another source of energy. Each of the considerations discussed above play a part in the ultimate decision, particularly the factor of availability. Consumers located in areas where natural gas is plentiful and bituminous coal deposits negligible will naturally select the former. Where two fuels are equally available, the less expensive will tend to be more widely employed. Electric utilities in Texas, for instance, were paying in 1948 61.3¢ per b.t.u. for fuel oil, only 7.9¢ for natural gas.[18] It is no wonder that utility plants in Texas used far more gas than fuel oil.

The most striking examples of interfuel competition based on price are to be found in areas totally dependent on imports of all fuels. In the northeastern United States (New York and New England), for example, bituminous coal and fuel oil have been close competitors in recent years, with competition based predominantly on comparative prices. At New York City during 1948–51 these were the delivered prices per b.t.u. for the two fuels: [19]

	7/1/48	*7/1/49*	*7/1/50*	*1/1/51*
Bituminous	34.4¢	34.8¢	33.9¢	33.7¢
Fuel Oil	58.7	25.9	32.2	33.6

The fall in price of fuel oil in 1949 had a serious effect on coal's market in the northeastern United States. This was reflected in market reports of that year:

BOSTON, July 7 (1949)—With a margin of around $4 a ton between the current price of coal and No. 6 fuel oil, in favor of the latter, and the oil trade anticipating another drop of 10 to 15 cents a barrel, it is not difficult to figure out what the numerous plants now operating on coal, but giving considerable consideration to converting, would do if coal prices are subjected to any further increase.

It is estimated that over 4,500,000 tons of bituminous coal have been lost to fuel oil in the New England states since December 1 [1948], when the price of oil started its decline. That is a substantial percentage of the total tonnage of bituminous coal used here last year, but worse than that, many of the conversions were the most valued accounts.[20]

The quickened pace of industrial and military activity which began in 1950 lifted oil prices from their 1949 low. Thereupon, market prospects for bituminous coal brightened. In mid-1951 an executive of a large coal firm declared: "A year ago residual fuel oil was taking a

18. The figures are for utility plants located in Dallas, prices as of July 1, 1948. Data from Federal Power Commission, as reported in National Coal Association, *Bituminous Coal Annual, 1951,* Table 67, p. 138.

19. Prices are for steam electric stations of public utilities. Ibid.

20. *Saward's Journal,* July 9, 1949, reprinted in Hearings, Senate Committee on Banking and Currency, *Economic Power of Labor Organizations,* p. 462.

good deal of business from coal along the Atlantic Coast and tide-water points. Today this picture is largely reversed and coal has regained most of this market and will regain more." [21]

An excellent indicator of the importance of price in interfuel competition is the number and location of electric-power plants equipped to burn two or more fuels. In 1947 there were 193 plants in operation of which 48 were "multi-fuel." [22] Power plants in the southwest (Texas, Louisiana, Arizona, and California) used natural gas. In the Appalachian area and the north central states (Pennsylvania, Ohio, Indiana, Michigan, and Wisconsin) bituminous coal was the principal fuel in use. All plants in Florida burned fuel oil. In states between the coal and natural gas regions both fuels were employed. In New England and New York coal and oil were utilized, natural gas being unavailable.[23]

Prices of the three fuels at points of production readily explain the preference for each according to geographic location of the consumer.

Table 36. Unit Value of Mineral Fuels at Point of Production, 1946–50

(CENTS PER MILLION B.T.U.)

Year	*Bituminous*	*Natural Gas*	*Crude Oil*
1946	13.1	4.9	24.3
1947	15.9	5.6	33.3
1948	19.0	6.0	44.8
1949	18.6	5.9	43.8
1950	18.5	6.0	43.8

Source: U. S. Bureau of Mines, as compiled in National Coal Association, *Bituminous Coal Annual, 1951,* Table 65, p. 138

In the regions, therefore, where natural gas is produced, it is clearly the least expensive of the competing fuels. In areas where it is not available for general use, its least expensive competitor will be employed—other things being the same.

21. National Coal Association, *Bituminous Coal Annual, 1951,* p. 99.

22. The distribution was as follows: bituminous coal, 103; coal and natural gas, 10; coal and fuel oil, 25; coal, oil, and gas, 3; oil, 10; gas, 26; lignite, 5; and lignite and gas, 1. Federal Power Commission, *Steam-Electric Construction Cost and Annual Production Expenses* (Washington [1948 ?]), pp. viii–x.

23. One factor which limits the extent to which natural gas can be used in industry is that domestic consumers are given preference on the available supply. Industrial users in some areas are, therefore, utilizing natural gas during off-peak periods and bituminous or fuel oil whenever gas is unavailable. Here again the preference for natural gas is primarily a matter of price.

Effects of Union Wage Bargains on Coal Consumption

While there has been a secular rise in other components of total cost—notably transportation charges—of bituminous coal, it is certain that increasing wage rates have had an important effect on selling prices. It can be acknowledged, too, that price increases contributed to a decline in the consumption of coal, relative to substitute fuels. It should be noted that the first proposition has been stated with no qualification, while the second is deliberately vague. For it is the principal conclusion of this chapter that selling price is only one of several factors entering into a consumer's decision to use a given fuel. It might be argued, in fact, that the more profound effect of union policy on sales has been the uncertainty of supply arising out of periodic work stoppages.

These reservations in mind, it seems that unionism has had some adverse affect on the market position of the bituminous industry. While this tends to confirm what most observers have long insisted, it must be stressed—even at the risk of excessive repetition—that there is no way to isolate the significance of the several factors which operate similarly to reduce total consumption of coal, relative to substitute fuels.

CHAPTER 9

An Appraisal of Union Policy and Its Effects

Broader Motivations of the UMW

Viewed superficially, union wage policy in the bituminous coal industry is "more and more—now." In this light the United Mine Workers seems to be a "business union," utterly devoted to the simple objective of ever improving wage rates and working conditions. In a broader perspective, however, the central motive of the UMW is more sophisticated. Union policy has been "security-oriented"—it insists that both the organization itself and its members be protected against the insecurity inherent in an industry highly sensitive to exogenous forces.

> Miners are haunted men. Their minds are vexed with the memories of bloody struggles for higher pay and for the preservation and growth of their labor union. Their thoughts are constantly troubled by insecurity of work, for they know that, although the calendar year contains 365 days, they have worked as little as [an average of] 142, and only nine times out of the last 25 years have they averaged more than 200 days a year. Their hearts grow weary repressing the importunate warnings of the dangers that lurk underground, which may at any time cut them off from their livelihood. Their families silently share their burdens.[1]

Though the prose in the quotation has a purplish tint, it cannot be denied that miners see themselves as a clan apart from the ordinary run of humanity. There is an unusual unity of outlook in economic matters among the leaders and members of the union. Memories of the strikes and violence and crushing poverty of the past have remained embedded in their minds. For them the union is not an instrument of social reform; it is their only protection against what they envision to be the implacable hostility of society in general and their employers in particular.[2] John L. Lewis put the issue to the miners

1. Madison, *American Labor Leaders*, p. 157.

2. A famous illustration of the attitude of mine operators in the antediluvian period is contained in a letter written in July 1902 by George F. Baer, president of the

in these flat terms: "I am content to serve my people. But remember, I am only as strong as you make me." [3] The miners, it can readily be seen, have done their part—and more.

The UMW's concern for security has several aspects, each of which has been adverted to in earlier chapters. First, the union has committed itself to assuring the income security of its members. This has been sought by constant effort to raise the real wages of the miners, while at the same time insisting on inflexibility downward of the prevailing level of wage rates. Second, it has labored to safeguard the structure of product prices against precipitous declines, since the union is fully aware of the close relationship between wages and prices. Third, it has undertaken to assure the income security of disabled and superannuated mine workers through the establishment of the miners' health and welfare fund.

Underlying most of these seemingly "bread-and-butter" objectives is an abiding depression consciousness in the union's top echelons. In its every move the UMW betrays its fear that, unless it is ever watchful, all of its good works will be washed away by an economic catastrophe. Thus, "excess capacity" must be eliminated lest the industry be left in a state of "overproduction." Pleas by the coal operators that demand for bituminous can be maintained only if unit producing costs are reduced are flatly rejected on two grounds: first, the miners' living standards must be maintained at any cost; and second, downward price-wage spirals have always in the past brought nothing but poverty to both mine owner and mineworker.

A similar logic underlies the UMW's uncompromising stand on uniformity in the structure of wage rates throughout the industry. No argument in favor of regional wage-rate differentials will be accepted by the union. When certain coal producers complain that their freight charges are higher, and that their survival depends on wage rates lower than those of their competitors, the union replies: "In every region there are mines which are farther from some markets than other firms, and there are some which are closer. The wage rate structure can't be perfect." Claims by operators that uniform wage

---

mine-owning Philadelphia and Reading Railroad. Addressing one of his stockholders who was worried about the violence attending a labor dispute, Baer wrote: "I beg of you not to be discouraged. The rights and interests of the laboring man will be protected and cared for—not by the labor agitators, but by the Christian gentlemen to whom God has given control of the property rights of the country and upon the successful management of which so much depends. . . . Pray earnestly that the right may triumph, always remembering that the Lord God Omnipotent still reigns and that His reign is one of law and order, and not of violence and crime." Coleman, *Men and Coal*, p. 70. Few of the coal producers at present would advocate dissolution of the UMW, let alone agree with Mr. Baer.

3. Sulzberger, *Sit down with John L. Lewis*, p. 163.

rates injure the higher-cost mines are rejected as an "old argument." Granting rate concessions to one field or even one mine requires that concessions be made to others—and the wage structure collapses as it did in bygone years.[4]

Nonunion mines and miners present a similar threat to the wage structure. Until recently at least, the UMW had dominion over nearly 90% of the mine labor force. There are evidences at this writing (early 1954) that an increasing number of nonunion mines are in operation. Should the output of these mines reach a volume that will affect selling prices in general, it is certain that the UMW will be forced to take strong counter-measures.[5]

The union's preoccupation with possible depressions has impelled it to devise its own remedies. The basic assumption of the UMW is that the interests of the operators are too disparate to allow united action by labor and management. The union, as a consequence, has arrogated to itself the responsibility for stabilization of the industry. This appears to mean that every effort must be exerted to preclude a general decline in wages, prices, and profits. The UMW has some strong, though not original, views on appropriate methods to deal with the problem.

In the autumn of 1948 the demand for bituminous coal began to recede from its postwar peaks. Speaking at the miners' biennial convention, John Lewis declared that if it became necessary, the union would stabilize the industry. The union would, he asserted, start a "share-the-work" plan to prevent "destructive competition," unemployment, and wage-rate reductions. "If evil days come on the industry again, you'll find the United Mine Workers moving in. And if there is only three days' work for this industry, then we'll have only three days' work. And if we are going to starve in this industry at any time, we'll just all starve together."[6] The threat was not idle; the "willing and able" clause that the union had thoughtfully inserted into the collective contract in 1947 made part-time operation possible and legal.

The plan was inaugurated tentatively on June 13, 1949, when the union began a stoppage described by Lewis as "a brief stabilizing period of inaction." There was an unmistakable implication that the

4. These views were stated to me in an interview with a union official, who expressed a wish that his identity be not revealed.

5. No data are presently available which reveal the extent of nonunion mining. If, as is likely, top officials in the industry know, they are keeping their own counsel. Reports in the press during the summer of 1952, however, indicated that union and nonunion miners in various sections of Ohio had been engaged in violence against one another.

6. *New York Times,* October 8, 1948, p. 21.

strike was aimed at reducing stockpiles above ground as a means of bolstering prices. Even the producers seemed pleased. Said one in the Pittsburgh area: "Many coal operators will be glad to see a shutdown. It has been all outgo and little income because of supply backlogs." [7] On June 25 the UMW proposed a uniform three-day week with a program of "divide-the-orders, share-the-work." [8] This was an attractive proposition to many firms, but it was resisted strenuously by the captive mines—all of which could sell a full week's output.[9] The views of the latter prevailed; the industry unanimously rejected the proposal.[10] Thereupon the UMW ordered its members to work a three-day week until further notice.[11]

The 1949 share-the-work, divide-the-orders plan was not a hastily contrived device to combat a particular crisis. In an earlier chapter it was commented that the concept of production control as a means of price maintenance has been a cherished ideal of the union since 1928.[12] The union was an ardent advocate of the NRA codes and of the Guffey Acts, and as recently as the summer of 1952 it was reported to have been sounding out the industry on a similar scheme.[13]

The unhappy truth is that the UMW is attempting to apply the "remedies" of monopoly (in the broad sense) to what it envisions are the central problems of the coal industry. What one noted economist once said about all labor unions certainly applies to the coal miners' union:

> . . . labor . . . distrusts all free-market ideas. It wants tariffs; it wants complete freedom from the Sherman Act; and, in fact, it wants employers who can fix their selling prices collusively too. American trade and industrial unionism makes sense only as part of a tight cartelization of industries where it is strong. It wants no competition from abroad and none at home, either in its own markets or in those of its employers. If employers will not or cannot police their product markets against chiselers,

7. Ibid., June 9, 1949, p. 1.
8. Ibid., June 26, 1949, p. 1.
9. Ibid., June 27, 1949, p. 1.
10. Ibid., June 29, 1949, p. 3.
11. Ibid., July 1, 1949, p. 3.
12. See the section on the union's political activities in Chap. 5.
13. "The scheme would operate in this way: Coal operators would agree with the union upon a standard uniform wage scale, to be paid, say, three or four days a week. If the miners worked four or five days the operators would pay a higher wage, in effect a penalty scale for those days. This, Mr. Lewis is represented as believing, by diverting production from full-time mining operations to mines which are working a short week, would tend to equalize work for all the coal diggers." Editorial in *New York Times,* commenting on a dispatch from Joseph A. Loftus, August 28, 1952, p. 22. Mr. Charles Killingsworth, public member and then Vice-Chairman of the Wage Stabilization Board, has expressed to me his view that the UMW "sold" its production control scheme for an additional 40¢ per day in the 1952 wage settlement. If this is true, the scheme was, of course, a mere bargaining lever.

unions will undertake that task themselves. Wage-fixing is price-fixing; labor monopoly means product monopoly even if employers compete effectively; and better wage bargains can be obtained from employers who do not compete with one another than those who do.[14]

Limitations on Union Policy

While it is important to understand the motivations and intent of the miners' union, it is equally important to recognize the boundaries within which the UMW must operate. There are two limitations on the union's wage policy: first, the level and elasticity of demand for the product; and second, the structure of the market and the behavior of the firms in the industry.

The greatest threat to the bituminous coal or any other industry is a secular decline in demand. If, because of a shift in preferences, consumers choose a substitute good, nothing but a campaign to regain their custom can prevent a steady decline in the activity of the industry producing the less desired product. If, then, the demand function shifts to the left, there will be a reduction in the number of men employed, the number of operating firms, product prices, and prices of productive services. In the face of diminishing demand for the product, attempts to rationalize the industry will amount to little more than a futile gesture.

There is nothing in the present policy of the UMW to suggest that it can prevent a secular decline in product demand. On the contrary, uncertainty of supply caused to consumers by the threat of strikes tends to reduce the level and increase the elasticity of demand for coal. Union officials are aware of what is happening, but they show no intent to revise their position. Said John Lewis in 1947:

> Our country still requires more coal than it did at any time in our history. That is a normal process, notwithstanding that perhaps 300 million tons, more or less, are displaced by the competition of natural gas, fuel oil, hydroelectric power and other substitutes. But that is all right. As long as those things are economically preferable, they should be encouraged. *But it does not make any difference whether we have 500,000 men in the coal mines or only 50,000 . . . they ought to be treated humanely; they ought to be paid a wage to protect their living standard. . . .*[15] [Italics added.]

The structure and behavior of the coal industry itself sets a further limit on union policy. Characterized by widespread ownership of coal lands, ease of entry, and rivalry among firms based primarily on sell-

14. H. C. Simons, *Economic Policy for a Free Society* (Chicago, University of Chicago Press, 1948), p. 103.

15. Hearings, House Committee on Education and Labor, *Welfare of Miners, 1,* 42–3.

ing price, the bituminous industry is highly competitive. Under those conditions it is virtually impossible for the union to get 100% acceptance of the union-enforced wage level. There will always, that is, tend to be a certain percentage of coal produced under nonunion conditions and at wage rates lower than in the organized sector of the industry. Moreover, the union will be faced constantly with the possibility that marginal firms will overtly or covertly abrogate their collective bargaining agreements and turn to operation on a nonunion basis. This threat will be greatly enhanced, of course, in periods when demand for coal has lessened and the marginal firms are being squeezed between high production costs and falling prices.

Experience indicates that the union does not need 100% acceptance of its wage scale to win its objectives. There is a "margin of tolerance" which can be permitted, though the precise upper and lower limits of the margin are indeterminate. It is indicative, however, that the union was at the peak of its power in the middle 1940's, although some 10% to 15% of bituminous output originated in nonunion mines. During the 1920's, on the other hand, less than half of the nation's coal production came from organized mines; under those conditions the union was virtually powerless.[16]

How would the UMW react were the margin of tolerance to be exceeded? If precedent is a reliable guide, it is likely that an effort would be made to cartelize the industry with public sanction. The intent would be to secure legally enforced floors under prices and wages. This was the purpose of the National Recovery and Bituminous Coal Conservation (Guffey) acts of the middle 1930's. Both statutes, it will be recalled, received vehement support from the UMW, both before and after their enactment.

Alternatives for Union Policy

Labor disputes in the bituminous industry have been frequent and spectacular in recent years. They have, as a result, provoked some bitter reactions among the many persons affected adversely. Some of those harmed have been content merely to fume. More sophisticated critics have, however, accused the miners' union of shortsightedness, of an excessive preoccupation with immediate gains. In this view union policy can only result in the long-run destruction of all job opportunities in the industry.

Whether one deplores, applauds, or is indifferent to the wage policy of the UMW, it is relevant to inquire whether the union has any

16. A similar situation prevails currently in the textile industry. The union's position in the northern mills is being undermined steadily by the flight of firms to the nonunion south.

viable alternatives to its present program. What is a "farsighted" wage policy? And would any policy other than the present one be consistent with the survival and internal stability of the union?

To many commentators a "responsible" union apparently is one which is "moderate" in its wage demands. Beyond that, responsible policy makers try to appraise the impact that their demands will have on product prices, sales, and the level of employment. Those who state the problem in more formal terms argue commonly that unionists must not drive wage rates above the level of competitive equilibrium.

As Arthur Ross has pointed out, this kind of reasoning assumes that there is a clear causal relationship among wage rates, unit costs, selling prices, and employment. The logic here is plausible, if not compelling. "Under the classical assumptions (constant production functions, consumer preferences and aggregate demand, and competitive product pricing), impulses would be transmitted reliably from one end to the other; a given change in wages would have a unique and predictable effect upon employment." [17]

This framework of analysis, built on a set of static assumptions, is of limited usefulness when it is applied to a dynamic situation. Actually, there is no clear connection between wage rates and total sales of the product. Price is only one factor, even though an important one, underlying a consumer's decision to use one fuel or another. What assurance is there, then, that some level of wage rates other than that prevailing will change significantly the quantity of coal demanded? As Ross has summarized the entire problem:

> Erratic fluctuations in business activity are clearly the most significant factor accounting for the indeterminateness of the wage bargain. Largely because of these fluctuations, wage rates and unit labor cost do not move together; unit labor cost and total unit cost frequently move in opposite directions; the effect of a change in cost upon price and the effect of a change in price upon employment cannot be foretold with any degree of confidence.[18]

Since a union leader must establish his policies under dynamic conditions, it is clear that, even if he desired, he could not appraise the economic effects of his actions. His first concern, consequently, will be to pursue a wage policy which will maximize the union's "institutional" objectives. In part, this means preserving internal cohesion within the organization by getting results for the working members. And in part, it means seeking to defend the union against present and future attacks by hostile employers or rival labor groups.

The policy of the United Mine Workers clearly follows this pat-

17. Ross, *Trade Union Wage Policy,* p. 80.
18. Ibid., p. 96.

tern. Through the years, and particularly since 1933, the leadership has maintained an undeviating course toward attaining the kind of environment it prefers to operate in. Successive wage bargains have included both worker-oriented and union-oriented provisions. There have been periods when the leadership has allowed a collective contract to extend unchanged beyond its termination date; typically this has happened when demand for coal has been slack. But it is important to note that the union has consistently kept a posture of aggressiveness; it has never indicated any intention of moderating its policies.

Behavior of this kind is susceptible of varying interpretations. On the one hand, for example, the aggressiveness of the leadership may be based on a correct interpretation of the wishes of the rank and file. The leaders may, on the other hand, feel so confident of their own ability to manipulate the attitudes of their constituents that membership sentiment is nothing more than a parameter in policy planning. Doubtless there are other hypotheses which are plausible.

The important point must be stated negatively: there is no evidence that the policy of the UMW or the vigor with which it is prosecuted is inconsistent with the internal stability of the group. Though the endurance and combativeness of the union has been tested severely in a number of grueling strikes in recent years, the dominance of the policy makers has not been challenged seriously since the middle twenties. On the contrary, the miners have supported each work stoppage almost to a man.

This does not mean that the membership would not support any less enthusiastically a more moderate policy. In view of the extent to which the policy-making officials dominate the organization and its organs of expression, it is likely that the leaders could "sell" their constituents on the need for a more cautious attitude. In Ross's words, the leaders "can elect to play either the role of militant fighters or that of cautious advisers. They must take care not to overrun the limits of discretion; but within these limits they have a wide range of choice." [19]

The union could, in sum, pursue a less aggressive policy. But such a policy would conflict with the official interpretation of what needs to be done. So long as the cohesion of the group is not threatened by the vigorous application of the policies deemed by the officials to be necessary, the policy makers can foresee no gains to be won by altering their course. In terms of the total situation, in short, the leaders of the miners' union need not concern themselves with policy alternatives.

19. Ibid., p. 41.

The Effects of Union Policy

The thrust of union policy is to control "prices via labor costs, [restricting] production as rapidly as consistent with decline of [the] membership by death and retirement and, while permitting some return to investors . . . to induce only as much employment as [the] original constituents [can] take care of without new members. If investors [dislike] high wages, they . . . like the high prices the [union] assures them by excluding low wage competitors." [20]

This policy yields results compatible not only with the interests of the United Mine Workers, but also with the interests of the low-cost and strategically situated firms in the industry. What is uncertain is the extent to which union policies of this kind have harmed the interests of other groups in the society. This depends largely, in turn, upon whether the economy is expanding or contracting.

Wage Rates and the Level of Employment

A general wage increase under conditions of full employment is certain to bring about an increase in the general price level. Since output cannot be expanded further, upward movements in costs and incomes must result in higher prices. There is no such certainty when wages rise during periods of underemployment. If prices lag behind the increase in wage rates, output and employment will expand; but if prices climb substantially, the may offset any tendency toward expansion which might have resulted from the wage increase.

An interesting question with regard to a "union-induced inflation" is the role of the monetary authorities. Their position is central, since a general rise in wages and prices will usually necessitate an increase in the quantity of money and credit. The dilemma of the central bankers is that they must choose between inflation and unemployment.

How will the Federal Reserve System react? On the basis of recent experience it seems true that "central banking authorities will always choose the inflationary path or if they should refuse to do so, business and labor will have them replaced by officials who will aid and abet the inflationary trend in the name of full employment." [21] It is not relevant here whether one accepts the flat assertion that business and labor can and will at all times exercise their power over the Chief Executive to assure themselves of a complaisant Board of Governors. It is more important to recognize that in a real sense the Reserve System may have no choice, that it may be unable to prevent a wage-

20. H. C. Simons, *Economic Policy for a Free Society,* p. 132.

21. W. A. Morton, "Trade Unionism, Full Employment and Inflation," *American Economic Review, 40* (1950), 29.

induced inflation. By the time, that is, that it becomes clear to the monetary authorities that a decision must be made, the new level of wages and prices will have been firmly established in the system. Under such conditions the central bankers will have no choice but to ratify the general wage increase by providing the greater amount of credit made necessary thereby.

WAGE RIGIDITY AND CYCLICAL ADJUSTMENTS

For reasons outlined in an earlier chapter, the policy of the United Mine Workers is to maintain wage rates during periods of depression. The effects of this policy, one common to most other labor organizations, have been a source of lively debate among economists. On the one hand, it is argued that wage-rate reductions will enable business firms to lower their selling prices, thus stimulating consumption. Moreover, cuts in wage rates, while causing a decline in individual incomes, will enlarge the total wage bill, since employers will be able to continue to employ the larger part of their work forces.

The contrary view is that wage cuts provide only a temporary respite to producers, since their lower costs will soon be offset by declining consumer spending. In addition, it is held likely that firms will defer new production and added expenditures on plant and equipment in the expectation that still lower wage costs will prevail later on.

There are further ramifications to both the views just sketched; for present purposes the simplified statements are sufficient. The experience of the coal industry indicates that unlimited flexibility of wage rates during the depression of the thirties failed to prevent a substantial reduction in employment. Furthermore, wage cuts opened the way for a downward spiral of selling prices, the principal effect of which was the virtual impoverishment of both miners and operators. While it is true that the downward rigidity in wage rates failed to halt both competitive price reductions and job losses in the recessions of 1937–38 and 1948–49, union-enforced "floors" under wages apparently softened the blow. To state the problem differently, wage cuts cannot counterbalance the decline in demand for labor which ensues from a drop in aggregate demand in the economy. But rigidity of rates can in some degree insulate firms against bankruptcy brought on by forces beyond their control.

WAGE CHANGES AND RESOURCE ALLOCATION

Under the classical assumptions of constant aggregate demand, consumer tastes and production functions, and competitive pricing of products, resource allocation is ideal when productive services are distributed among competing uses in precise accordance with the wishes of consumers. It follows that buyers will get just those commodities and services they want in just the amounts they want them. In addition, profit-maximizing firms will be operating at the least-cost point on both their short- and long-run average cost functions. To the extent, furthermore, that there is "perfect" competition in the hire of productive services, the rewards of the latter will be measured by the value of their marginal product.

The stated assumptions are, of course, an abstraction from reality; in the real world there are certain to be imperfections of one sort or another. Even under nonunion conditions, for example, wage rates may be below the level of competitive equilibrium because of labor immobility or imperfect knowledge, among other things. For reasons such as these, unionism may produce a better or poorer allocation of resources.

Under conditions of constant aggregate demand a successful attempt by unions to raise wage rates above the level which market forces would establish would result in higher product prices, lower output, and a lower volume of employment. Consumer choices would perforce be altered from the pattern freely selected. And the resources displaced by union action would be forced into the nonunion sector of the economy, depressing factor prices in the unorganized areas.

These results appear to constitute a significant deviation from the ideal. Suppose, however, that there are important differences in costs among the firms in the unionized industries. Upward pressure on wage rates may compel the high-cost concerns to effect offsetting economies in other cost items, while maintaining the same level of employment. Some producers will succeed; others who cannot will be squeezed out of the industry, freeing resources which can be employed more effectively elsewhere. In such a situation, even if aggregate demand is unchanged, allocation may be improved.

Assuming an expansion in total spending, union pressure on wage rates may have even greater favorable effects on resource allocation. The wage policy of the UMW, for example, has probably improved allocation by hastening the liquidation of the persistent excess of labor attached to the industry. This has come about in two ways. First, the increase in the earnings of employed miners has enabled them to finance the training of their children for jobs in other industries. Second, the union-imposed rigidity of wage rates downward and flexibility upward has stimulated appreciably the substitution of capital for labor. Union floors under wage rates are a deterrent to price wars and thus help to stabilize the industry, lessening entrepreneurial risks and encouraging a greater volume of investment. And rising labor costs, particularly in an industry faced with an elastic demand function over the long-run, will surely stimulate efforts to effect counterbalancing economies.

It must be stressed, however, that it is rising aggregate demand which most effectively translates the effects of union policy into a general increase in consumer welfare. Since the demand for coal is a derived demand, expanding business activity makes possible a concomitant increase in wage rates and profits. Technological change becomes feasible financially. Moreover, the availability of alternative employment opportunities in expanding industries permits the resources displaced from the coal industry to move to areas where they are more highly valued.

An important conclusion which emerges from this brief analysis is that vigorous unionism, as epitomized by the UMW, need not result in misallocation of resources in either a static or a dynamic economy. Indeed, that kind of unionism may improve allocation under both situations, though probably more so under the latter than the former.

WAGE CHANGES AND INCOME DISTRIBUTION

What effects flow from successful attempts by labor unions to push up wage rates faster than productivity of labor is rising? According to the assumptions made, the distribution of income may be altered, the rate of capital accumulation retarded, the general price level inflated, or a combination of these results brought about. But if the assumptions are changed, none of these things or even their opposites might occur.

Because the functional distribution of income bears on the pattern of spending in the system, it is important to determine whether union pressure can increase the share of national income paid out to workers. Some valuable inferences can be drawn but it is well to remember that conclusions differ with a change in assumptions. Suppose, for example, that union pressure on wage rates induces employers to substitute capital for labor. Since in this situation the demand for labor is elastic, the subsequent reduction in employment in the industry will reduce total wages paid out. The wage-income ratio will decline. But assume, on the other hand, that the firms adjust to higher wage rates by raising prices. If the demand for the product is elastic, total receipts of the firm will fall. Under these conditions the proportion of wages to income will be greater as a result of union action. In the short run, then, it appears that union-induced wage changes can alter the distribution of income, though not always in favor of the workers' earnings at the expense of entrepreneurial returns.

Recent studies indicate that in the long-run union pressure on wages has not changed significantly the share of income paid to labor.[22] This does not prove that unionism has no influence on the functional distribution of income, for there is no way to determine what the pattern of distribution would have been had there been no unions. Nonetheless, as Paul Sultan has summarized his findings: ". . . *whatever the impact of union pressure,* it has not served to increase the distributive share going to labor in those industries which are highly unionized, relative to those industries which are not."[23]

While it is doubtful that unionization can change appreciably the functional distribution of income, it seems probable union pressures can affect the distribution between organized and unorganized workers. If higher wage rates result in capital substitution or upward price adjustments, employment in the organized industries will decline, though the employment effect depends in part on the elasticity of demand for the product with respect to price. Money and real wages of unionized workers will advance relative to unorganized workers, since wage rates in the nonunion sector will lag behind those of unionists, displaced workers seeking new jobs will tend to depress wage rates in nonunion industries, and unorganized employees will be faced with the higher prices of union-made products. Workers in unionized industries can, then, raise their level of real wages by pushing up their money wage rates, so long as their wages rise faster than those in the nonunion sector of the economy. But as the number of organized workers in the economy increases, this possibility diminishes.

WAGE CHANGES AND CAPITAL ACCUMULATION

In an earlier chapter it was stated that expenditures on new mining equipment were stimulated by rising wage rates. If this were uniformly the case, union pressure on the level of wage rates might reduce employment in a firm or industry but tend to raise employment in the economy as a whole. To the extent, moreover, that rapid accumulation of capital is consistent with expanding living standards, union policy may promote long-run growth in income and employment.

Under certain conditions, however, wage increases may retard capital accumulation. This would occur, for example, in industries the demand for whose products is elastic with respect to price. Firms would find themselves squeezed between higher costs and falling receipts. Profit margins would be narrowed, thereby reducing the amount of money available for purchases of new equipment.

22. See, for instance, H. M. Levinson, *Unionism, Wage Trends, and Income Distribution, 1914–1947,* Michigan Business Studies, *10,* No. 4 (Ann Arbor, University of Michigan Press, 1951); P. E. Sultan, "Unionism and Wage-Income Ratios, 1929–1951," *Review of Economics and Statistics, 36* (1954), 67–73.

23. Ibid., p. 73.

In a society bent on technological progress and capital accumulation almost at any price, situations of the latter type will be a source of general consternation. On the other hand, "a slower rate of capital accumulation may be accompanied by greater economic stability, and it may be argued that this possibility makes the slowing-up of economic expansion well worth while." [24]

Conclusion

The attempt has been made here to appraise the effects of policies like those of the United Mine Workers in the perspective of the economy as a whole. Necessarily the analysis has been sketchy. However, the treatment suffices to demonstrate the difficulties inherent in an evaluation of the impact of unionism on the economy and its component parts.

The clear implication emerges that no positive case can be made for or against unionism on economic grounds. This suggests that efforts to control union activities must be directed more against the political than the economic effects of unions. Of this economists are becoming increasingly aware, although few have gone beyond generalized worrying about the relationship of big unionism to allegedly nascent syndicalism in our society. Concentration of control is a valid concern, but cannot be the touchstone for public policy toward labor organizations. Policy must be framed in broader terms. As C. E. Lindblom has commented, "The first and perhaps primary requirement of reform is that it maintain and enlarge freedom. . . . The second, that it be consistent with economic efficiency. The third, that it protect the individual from economic insecurity." [25]

24. L. G. Reynolds, *Labor Economics and Labor Relations* (New York, Prentice-Hall, 1949), p. 442.

25. C. E. Lindblom, *Unions and Capitalism* (New Haven, Yale University Press, 1949), p. 237.

Appendix

Table I-1. Average Producing Costs per Ton of Commercial Bituminous Mines in Price Area I, All Types of Mines, 1945

District	Mine Labor Cost	Mine Supply Cost	Other Mine Expenses	Other Oper. Charges	Total Prod. Cost
1	$2.1902	$.4731	$.0954	$.4221	$3.1808
2	1.9195	.4329	.0934	.4979	2.9437
3	1.5932	.4790	.0683	.4213	2.5618
4	1.4811	.5318	.0669	.4146	2.4944
6	1.8000	.4082	.0815	.4663	2.7560
7	2.3651	.5347	.0386	.4562	2.3946
8	1.9784	.4751	.0536	.4315	2.9386

Note: District 5 figures are omitted to avoid disclosing information regarding individual operations.

Source: Office of Temporary Controls, *Preliminary Survey of Operating Data for Commercial Bituminous Coal Mines,* OPA Economic Data Series No. 2 (Washington, Government Printing Office, 1946), p. 7

Table I-2. Average Producing Costs per Ton for Commercial Bituminous Mines in Price Area I, Machine-loaded Mines, 1945

District	Mine Labor Cost	Mine Supply Cost	Other Mine Expenses	Other Oper. Charges	Total Prod. Cost
1	$2.2268	$.6088	$.0726	$.4056	$3.3138
2	1.9841	.5896	.0313	.4994	3.1044
3	1.5754	.5325	.0345	.4082	2.5506
4	1.7233	.5532	.0450	.3596	2.6811
6	2.1703	.4742	.0225	.3211	2.9881
7	2.1037	.6238	.0422	.4759	3.2456
8	1.7882	.5228	.0641	.4516	2.8267

Source: ibid., p. 19

Table I-3. Average Producing Costs per Ton in Commercial Bituminous Mines in Price Area I, Strip Mines, 1945

District	Mine Labor Cost	Mine Supply Cost	Other Mine Expenses	Other Oper. Charges	Total Prod. Cost
1	$1.5017	$.2757	$.2651	$.6448	$2.6873
2	1.2722	.3384	.2605	.6445	2.5156
3	1.2535	.3674	.1893	.5973	2.4075
4	.8699	.5638	.1080	.5080	2.0497
6	.8869	.4603	.2395	.8443	2.4310
7	2.3861	.6332	.0225	.3561	3.3979
8	1.8054	.3630	.1300	.4223	2.7207

Source: ibid., p. 25

Table I-4. Average Producing Costs per Ton in Commercial Bituminous Mines in Price Area I, Hand-loaded Mines, 1945

District	Mine Labor Cost	Mine Supply Cost	Other Mine Expenses	Other Oper. Charges	Total Prod. Cost
1	$2.3540	$.4772	$.0630	$.3719	$3.2361
2	2.2563	.3542	.0458	.4077	3.0640
3	1.9810	.3446	.1068	.3187	2.7511
4	2.1607	.3744	.0414	.3750	2.9515
6	1.9836	.3172	.0450	.3830	2.7288
7	2.4602	.4986	.0378	.4525	3.4491
8	2.1623	.4334	.0414	.4129	3.0500

Source: ibid., p. 13

Table I-5. Percentage of Bituminous Coal Produced in Price Area I, by Depth of Seam, Underground Mines, 1945

District	Under 2 Ft.	2–4 Ft.	4–6 Ft.	6–8 Ft.	Over 8 Ft.
1	1.0%	71.3%	26.5%	0.6%	0.6%
2	0.5	5.6	32.6	49.8	11.5
3	—	10.0	17.1	58.8	14.1
4	—	18.8	79.6	1.6	—
6	—	0.3	99.7	—	—
7	—	49.8	30.9	14.7	4.6
8	—	47.5	43.0	8.8	0.7

Source: W. H. Young and R. L. Anderson, *Thickness of Bituminous Coal and Lignite Seams Mined in U. S. in 1945*, Information Circular 7442 (U. S. Bureau of Mines, 1947), p. 3

Table 1-6. Average Output per Man per Day in Price Area I, Commercial Bituminous Mines, 1945

District	*Average Tons per Day*	*Range for Component Fields*
1	4.97	2.36– 7.29
2	5.69	4.98–10.24
3	7.29	2.95–10.09
4	7.27	2.93–21.55
6	5.86	4.61– 9.27
7	4.53	3.57– 5.15
8	4.88	2.29– 6.52

Source: U. S. Bureau of Mines, *Minerals Yearbook* (1946), pp. 326–39, and Information Circular 7442, *Thickness of Bituminous Coal Seams . . . Mined in U. S. in 1945*, p. 5

Table 1-7. Comparison of Various Kinds of Loading Devices

Type of Machine	*Horsepower*	*Daily Capacity*
Pit-car Loader	1–5	15–25 (tons)
Hand-loaded Face Conveyor	5–30	50–300
Duckbill	15–30	50–300
Scraper	7½–25	50–250
Mobile Loader	22½–50	100–800

Source: U. S. WPA National Research Project, *Mechanization, Employment and Output per Man in Bituminous Coal Mining*, Report E-9 (2 vols. 1939), p. 116

Table 1-8. Labor Output and Labor Requirements in Selected Bituminous Mines in Periods of Hand and Machine Loading, by Type of Loader *

		TONS PRODUCED PER MAN-HOUR			NO. OF MEN REQUIRED TO MAINTAIN 1,000 TONS OF HOURLY CAPACITY		
Type of Loader	*No. of Mines*	*Hand Load. Period*	*Mech. Load. Period*	*Pct. Inc.*	*Hand Load. Period*	*Mech. Load. Period*	*Pct. of Decrease*
Scraper	5	.396	0.695	75.5	2,525	1,439	43.0
Mobile	39	.736	1.116	51.6	1,359	896	34.1
Duckbill	5	.777	.954	22.8	1,287	1,048	18.6
Pit-car	34	.698	.833	19.3	1,433	1,200	16.3
Hand Face Conveyer	21	.435	.503	15.6	2,299	1,988	13.5

* Computed from data of the U. S. Bureau of Mines. In general, mines were selected which showed that in two years or more 75% or more of total produc-

tion had been loaded by one type of equipment. Figures for the mechanical-loading period represent average performance of from two to four years under mechanical loading ranging from 75% to 100% of total mine output. The hand-loading period covered was an average of three or four years under representative hand-loading conditions. Mines could not, of course, be included if there was no hand-loading period; i.e., mines that opened up completely mechanized [and those] for which it was difficult to get a representative hand-loading period were omitted. Mines also were omitted that showed indications of inaccurate reporting. In order to retain the same relative weights of individual mines under both mechanical and hand-loading conditions, so as to reflect the composite change in productivity unaffected by changes in the relative magnitude of production between the two periods, the average for each mine was weighted on the basis of its relative significance (as measured by man-hours worked) during the period of mechanical loading covered.

Source: Coal Age, 45 (1940), p. 51

Table 4-1. Financial Condition of United Mine Workers, 1890–98

Year	*Income*	*Expenditures*	*Surplus*
1890	$54,314.33	$38,583.42	$15,730.91
1891	70,025.78	63,430.38	6,595.40
1892	41,927.31	31,559.09	10,368.22
1893	30,928.84	25,365.76	5,563.08
1894	28,847.06	28,350.23	496.83
1895	15,280.08	14,124.31	1,155.77
1896	11,434.45	10,851.52	582.93
1897	39,165.90	28,353.72	10,812.18
1898	60,609.33	37,719.02	22,890.31

Source: C. B. Fowler, *Collective Bargaining in the Bituminous Coal Industry* (New York, Prentice-Hall, 1927), p. 43

Table 4-2. Basic Tonnage Rates by Districts in Appalachian Area, 1935 and 1941

	TONNAGE RATES PER 2000 LBS. RUN-OF-MINE COAL		
District	*1935*	*1941*	*Increase*
Western Pennsylvania *			
Pick Mining, Thin Vein	$.89	$1.10	$.21
Machine Loading, Thin Vein	.68	.87	.19
Cutting, Shortwall Machine	.10	.12	.02

* Thick-vein rates are omitted since all other district rates are based on the western Pennsylvania thin-vein rates.

	TONNAGE RATES PER 2000 LBS. RUN-OF-MINE COAL		
District	*1935*	*1941*	*Increase*
Central Pennsylvania			
Pick Mining	$.89	$1.10	$.21
Machine Loading	.68	.87	.19
Cutting, Shortwall Machine	.10	.12	.02
So. Somerset County, Pa.			
Pick Mining	.89	1.10	.21
Machine Loading	.68	.87	.19
Cutting, Shortwall Machine	.10	.12	.02
Connellsville, Pa.			
Pick Mining	.75	.96	.21
Machine Loading	.56	.75	.19
Cutting, Shortwall Machine	.08	.11	.03
Westmoreland-Greensburg, Pa.			
Pick Mining	.75	1.05	.30
Machine Loading	.56	.83	.19
Cutting, Shortwall Machine	.08	.11	.03
Thick Vein Freeport, Pa.			
Pick Mining	.84	1.05	.21
Machine Loading	.64	.83	.19
Cutting, Shortwall Machine	.09	.11	.02
Northern West Va.			
Pick Mining	.75	.96	.21
Machine Loading	.585	.775	.19
Cutting, Shortwall Machine	.085	.105	.02
Ohio and Panhandle of No. West Va.			
Pick Mining	.89	1.10	.21
Machine Loading	.68	.87	.19
Cutting, Shortwall Machine	.10	.12	.02
Maryland and Upper Potomac			
Pick Mining	.812	—	—
Machine Loading	.68	.80	.19
Cutting, Shortwall Machine	.10	.12	.02
Bakerstown Seam			
Pick Mining	.87	1.08	.21
Machine Loading	.73	.92	.19
Cutting, Shortwall Machine	.10	.12	.02
Waynesburg Seam			
Pick mining	.87	1.08	.21
Kanawha			
Machine Loading	.68	.87	.19
Cutting, Shortwall Machine	.10	.12	.02
Machine Loading	.582	.772	.19
Cutting, Shortwall Machine	.09	.11	.02

Table 4-2. (continued) Basic Tonnage Rates by Districts in Appalachian Area, 1935 and 1941

District	TONNAGE RATES PER 2000 LBS. RUN-OF-MINE COAL		
	1935	*1941*	*Increase*
Logan			
Machine Loading	$.492	$.682	$.19
Cutting, Shortwall Machine	.072	.092	.02
Williamson			
Machine Loading	.518	.708	.19
Cutting, Shortwall Machine	.076	.096	.02
Big Sandy—Elkhorn			
Machine Loading	.625	.815	.19
Cutting, Shortwall Machine	.10	.12	.02
Hazard			
Machine Loading	.562	.752	.19
Cutting, Shortwall Machine	.10	.12	.02
Harlan			
Machine Loading	.57	.76	.19
Cutting, Shortwall Machine	.09	.11	.02
Virginia			
Machine Loading	.568	.758	.19
Cutting, Shortwall Machine	.09	.11	.02
Southern Appalachian			
Machine Loading	.59	.78	.19
Cutting, Shortwall Machine	.10	.12	.02
New River			
Machine Loading	.602	.792	.19
Cutting, Shortwall Machine	.10	.12	.02
Pocohontas—Tug River			
Machine Loading	.517	.707	.19
Cutting, Shortwall Machine	.065	.085	.02
Winding Gulf			
Machine Loading	.544	.734	.19
Cutting, Shortwall Machine	.09	.11	.02
Greenbrier			
Machine Loading	.552	.742	.19
Cutting, Shortwall Machine	.075	.095	.02

Source: M. Coleman, *Men and Coal* (New York, Farrar & Rinehart, 1943), Appendix A, pp. 324–7; *Wage Agreements, 1935–1936,* UMW[1935?], pp. 33–6

BIBLIOGRAPHY

Alinsky, S., *John L. Lewis,* New York, Putnam, 1949.

American Mining Congress, "Coal Mine Modernization," *Yearbook,* 1943.

Appalachian Coals, Inc. v. U. S., *U. S. Reports, 288* (1933).

Backman, J., *Bituminous Coal Wages, Profits, and Productivity* [Washington?], Southern Coal Producers' Association, 1950.

Baker, R. H., *The National Bituminous Coal Commission,* Baltimore, Johns Hopkins University Press, 1941.

Barger, H., and Schurr, S. H., *The Mining Industries, 1899–1939. A Study of Output, Employment, and Productivity,* New York, National Bureau of Economic Research, 1944.

Bargeron, C., "Better Jobs for Better Miners," *Nation's Business, 33* (October 1945).

Baruch, B. M., "Coal. Reply with Rejoinder," *New Republic, 115* (December 30, 1946).

Bateman, A. M., *Economic Mineral Deposits,* 2d ed. New York, John Wiley, 1950.

Berquist, F. E., and Associates, *Economic Survey of the Bituminous Coal Industry under Free Competition and Code Regulation,* Work Materials No. 69, Washington, U. S. National Recovery Administration, Division of Review, 1936.

Bloch, L., *Coal Miners' Insecurity,* New York, Russell Sage, 1922.

———, *Labor Agreements in Coal Mines,* New York, Russell Sage, 1931.

Bowden, W. L., "Changing Status of Bituminous Coal Miners, 1937–1946," U. S. Department of Labor, *Monthly Labor Review, 63* (1946).

Business Week.

Carnes, C., *John L. Lewis. Leader of Labor,* New York, McLeod, 1936.

Carter v. Carter Coal Co., *U. S. Reports, 298* (1936).

Chamberlain, J., "The Special Case of John L. Lewis," *Fortune, 28* (1943).

Chamberlain, N. W., *Collective Bargaining,* New York, McGraw-Hill, 1951.

———, *The Union Challenge to Management Control,* New York, Harper, 1948.

Coal Age.

Coleman, M., *Men and Coal,* New York, Farrar & Rinehart, 1943.

The Consumers' Counsel and the National Bituminous Coal Commission, Washington, Committee on Public Administration, 1949.

Cowan, D. R. G., *More Capital Equipment. Coal's Foremost Economic Need,* Washington, National Coal Association, 1948 (pamphlet).

Dahl, R. A., and Lindblom, C. E., *Politics, Economics, and Welfare,* New York, Harper, 1953.

Devine, E. T., *Coal. Economic Problems of Mining,* Bloomington, Indiana, American Review Service Press, 1925.

Douglas, P. H., *Real Wages in the United States,* Boston, Houghton Mifflin, 1930.

Dron, R. W., *Economics of Coal Mining,* New York, Longmans, 1928.

"The Economics of Coal," *The Economist,* London, September 17, 1946.

Emmet, B., *Labor Relations in the Fairmont, West Virginia Bituminous Coal Field,* Washington, U. S. Department of the Interior, 1924.

Evans, C., *History of the United Mine Workers of America,* 2 vols. Indianapolis [United Mine Workers of America, 1918?].

Fisher, W. E., "Bituminous Coal," in *How Collective Bargaining Works,* Dickinson, Z., ed., New York, Twentieth Century Fund, 1942.

———, *Collective Bargaining in the Bituminous Coal Industry. An Appraisal,* Philadelphia, University of Pennsylvania Press, 1948 (pamphlet).

———, *Economic Consequences of the Seven-Hour Day and Wage Change in the Bituminous Coal Industry,* Philadelphia, University of Pennsylvania Press, 1939.

———, "Union Wage and Hour Policies and Employment," *American Economic Review, 30* (1940).

———, "Wage Rates in the Bituminous Coal Industry," *Report of the United States Coal Commission,* Pt. 3 (1925).

———, and Bezanson, A., *Wage Rates and Working Time in the Bituminous Coal Industry, 1912–1922,* Philadelphia, University of Pennsylvania Press, 1932.

Fortune, "Coal," *35* (March and April 1947).

Fortune, "Coal. The 'Pitt Consol' Adventure," *36* (July 1947).

Fowler, C. B., *Collective Bargaining in the Bituminous Coal Industry,* New York, Prentice-Hall, 1927.

Friendly, A., "What Does Lewis Want?," *Nation, 162* (1946).

Fritz, W. G., and Veenstra, T. A., "Major Economic Tendencies in the Bituminous Coal Industry," *Quarterly Journal of Economics, 50* (1936).

———, *Regional Shifts in the Bituminous Coal Industry,* Pittsburgh, University of Pittsburgh Press, 1935.

Galloway, G. B., *Industrial Planning Under the Codes,* New York, Harper, 1935.

Glasser, C., "Union Wage Policy in Bituminous Coal," *Industrial and Labor Relations Review,* July 1948.

Gluck, E., *John Mitchell, Miner,* New York, John Day, 1929.

Goodrich, C. L., *The Miners' Freedom,* Francestown, N. H., Marshall Jones, 1925.

Graham, H. D., *The Economics of Strip Coal Mining,* Economic Paper No. 11, Washington, U. S. Bureau of Mines, 1931.

Haley, J. W., "Minimum Wages in the Bituminous Coal Industry," *Congressional Digest, 24* (1945).

Hamilton, W. H., "Coal and the Economy—A Demurrer," *Yale Law Journal, 50* (February 1941).
———, and Wright, H. R., *The Case of Bituminous Coal,* New York, Macmillan, 1926.
Hapgood, P., *In Non-Union Mines. The Diary of a Coal Digger,* New York, Bureau of Industrial Research, 1922.
Hardman, J. B. S., ed., *American Labor Dynamics,* New York, Harcourt, Brace, 1928.
Harris, H., *Labor's Civil War,* New York, Knopf, 1940.
Hinrichs, A. F., *The United Mine Workers and the Non-Union Coal Fields,* New York, Columbia University Press, 1923.
Hunt, E. E., Tryon, F. G., and Willitts, J. H., *What the Coal Commission Found,* Baltimore, Williams and Wilkins, 1925.
Hutchison, K., "Competitors of Coal," *Nation, 163* (1946).
Ickes, H. L., "Crisis in Coal," *Collier's, 112* (September 1943).
International Labor Office, *Reports of Coal Mines Committee,* Meetings of 1947–50.
Interstate Commerce Commission, "In Re: Increase in Freight Rates and Charges, 1935," Brief on Behalf of the National Bituminous Coal Commission and Consumers' Counsel, 1935, Pt. 1.
Island Creek Coal Company, *Annual Reports.*
Josephson, M., *Sidney Hillman. Statesman of American Labor,* New York, Doubleday, 1952.
Keystone Coal Buyers' Annual.
Lerner, D., and Lasswell, H. D., *The Policy Sciences,* Stanford, Calif., Stanford University Press, 1951.
Lester, R. A., *Labor and Industrial Relations,* New York, Macmillan, 1951.
———, "Southern Wage Differences," *Southern Economic Journal,* April 1947.
———, and Shister, J., *Insights into Labor Issues,* New York, Macmillan, 1948.
Lewis, J. L., "The Effect of Moderate and Gradual Wage Increases on Prices and Living Costs," *The Annalist, 50* (November 12, 1937).
———, *The Miners' Fight for American Standards,* Indianapolis, Ball Publishing Co., 1925.
———, "United Mine Workers' Demands," *Reference Shelf,* No. 4 (1946).
Lewis, W. A., *Overhead Cost,* London, Allen & Unwin, 1949.
Lubell, S., *The Future of American Politics,* New York, Harper, 1952.
Lubin, I., *Miners' Wages and the Cost of Coal,* New York, McGraw-Hill, 1924.
MacDonald, D. J., and Lynch, E. A., *Coal and Unionism,* Silver Spring, Md., Cornelius Printing Co., 1939.
Madison, C. A., *American Labor Leaders,* New York, Harper, 1950.
Menefee, S. C., "Why They Follow John L. Lewis," *Nation, 156* (April 3, 1943).
Miller, J. P., "Pricing of Bituminous Coal. Some International Compari-

sons," in *Public Policy,* Friedrich, C. J., and Mason, E. S., eds., Cambridge, Harvard University Press, 1940, *1.*
Mills, C. W., *The New Men of Power,* New York, Harcourt, Brace, 1948.
Mitchell, J., *Organized Labor,* Philadelphia, American Book and Bible House, 1903.
Moody's Manual.
Mansfield, H. C., *The Lake Cargo Coal Rate Controversy,* New York, Columbia University Press, 1932.
Moore, E. S., *Coal,* New York, John Wiley, 1922.
Murray, P., and Cooke, M. L., *Organized Labor and Production,* New York, Harper, 1940.
National Bureau of Economic Research, *Report of the Committee on Prices in the Bituminous Coal Industry,* New York, 1939.
National Coal Association, *Bituminous Coal Annuals,* 1949–51.
———, *Bituminous Coal Data, 1935–48* and *1951.*
National Industrial Conference Board, *The Competitive Position of Coal in the United States,* New York, 1931.
National War Labor Board, *In Re: National Bituminous Coal Conference and United Mine Workers of America,* Case No. 111–14875–D, 1945 (Opinion by G. W. Taylor, Chairman).
"New Coal Mining Tools," *Fortune, 38* (June 1950).
New York Times.
Parker, G. L., *The Coal Industry. A Study in Social Control,* Washington, American Council on Public Affairs, 1940.
Poor's Manual.
Power Survey Commission, *Power in New England,* Boston, New England Council, 1948.
Raskin, A. H., "The Secret of John L. Lewis' Great Power," *New York Times,* Magazine Section, October 5, 1952.
Reynolds, L. G., "Cutthroat Competition," *American Economic Review, 30* (1940).
———, "Toward a Short-run Theory of Wages," *American Economic Review, 38* (June 1948).
Ross, A. M., *Trade Union Wage Policy,* Berkeley, University of California Press, 1948.
Ross, M. H., *Machine Age in the Hills,* New York, Macmillan, 1933.
Rostow, E. V., "Bituminous Coal and the Public Interest," *Yale Law Journal, 50* (1941).
Rowe, J. W. F., *Wages in the Coal Industry,* London, King, 1923.
Ryan, J. T., "The Future of the Bituminous Coal Industry," *Harvard Business Review, 14* (1936).
Shurick, A. T., *The Coal Industry,* Boston, Little, Brown, 1924.
———, *Coal Mining Costs,* New York, McGraw-Hill, 1922.
Slichter, S. H., *Union Policies and Industrial Management,* Washington, Brookings, 1941.
Simons, H. C., *Economic Policy for a Free Society,* Chicago, University of Chicago Press, 1948.

Smart, R. C., *The Economics of the Coal Industry,* London, King, 1930.

Stewart, W., *Mines, Machines, and Men,* London, King, 1935.

Suffern, A. E., *The Coal Miners' Struggle for Industrial Status,* New York, Macmillan, 1926.

———, *Conciliation and Arbitration in the Coal Industry of America,* Boston, Houghton Mifflin, 1915.

Tarboux, J. G., *Electric Power Equipment,* 3d ed. New York, McGraw-Hill, 1946.

Thomas, I., *Coal in the New Era,* New York, Putnam, 1934.

Truman, D. B., *The Governmental Process,* New York, Knopf, 1951.

Tryon, F. G., "Effect of Competitive Conditions on Labor Relations in Coal Mining," *Annals* of the American Academy of Political and Social Science, *111* (1924).

———, *The Trend of Coal Demand,* Columbus, Ohio State University Press, 1929.

———, and others, *Mineral Economics,* New York, McGraw-Hill, 1932.

U. S. Bureau of the Census, *Decennial Censuses of the Population.*

U. S. Bureau of Internal Revenue, *Statistics of Income.*

U. S. Bureau of Labor Statistics, *Productivity and Unit Labor Cost in Bituminous Coal Mining, 1937–1948,* 1948.

U. S. Bureau of Labor Statistics, *Wage Structure in Bituminous Coal Mining, Fall of 1945,* Bulletin No. 867, 1946.

U. S. Bureau of Mines, *Technical Bulletins* and *Information Circulars.*

U. S. Coal Commission, *Report on the Effect of Irregular Operation on the Unit Cost of Production,* Washington, 1925.

U. S. Department of Commerce, *Survey of Current Business* and *National Income Supplement.*

U. S. Department of Labor, *Monthly Labor Review.*

U. S. Federal Emergency Relief Administration, *Bituminous Coal Industry with a Survey of Competing Fuels,* 1935 (photoprint).

U. S. Federal Mediation and Conciliation Service, *Report to the President,* 1950.

U. S. Federal Power Commission, *Natural Gas Investigation,* Docket G-580, 1948.

U. S. Geodetic Survey, *The Coal Fields of the United States,* Professional Paper 100, 1929.

U. S. House of Representatives, Committee on Education and Labor, *Hearings on Amendments to the National Labor Relations Act,* 80th Congress, 1st Session, 1947.

———, *Hearings on Welfare of Miners,* 80th Congress, 1st Session, 1947.

U. S. Interstate Commerce Commission, Bureau of Transportation Economics and Statistics, *Study of Railroad Motive Power,* Statement 5025, 1950.

———, Bureau of Statistics, *Freight Rates on Bituminous Coal, with Index Numbers, 1929–1940,* Statement 413, 1941.

———, *Statistics of Railways of the United States.*

U. S. National Labor Relations Board, Division of Economic Research,

Effect of Labor Relations in the Bituminous Coal Industry upon Interstate Commerce, Bulletin No. 2, 1938.

———, *Written Trade Agreements in Collective Bargaining,* Bulletin No. 4, 1939.

U. S. National Recovery Administration, *Report of Mining Engineers on the . . . Shorter Work Day and Work Week,* 1934 (photoprint).

U. S. Office of Temporary Controls, *Preliminary Survey of Operating Data of Commercial Bituminous Coal Mines,* Economic Data Series No. 2, 1946.

———, *Survey of Commercial Bituminous Coal Mines,* Economic Data Series No. 15, 1946.

U. S. Senate, Committee on Banking and Currency, *Hearings on Economic Power of Labor Organizations,* 81st Congress, 1st Session, 1949.

———, Committee on Interstate Commerce, *Hearings on Stabilization of the Bituminous Coal Industry,* 74th Congress, 1st Session, 1935, and 75th Congress, 1st Session, 1937.

———, Committee on Labor and Public Welfare, *Hearings on Causes of Unemployment in the Coal Industry,* 81st Congress, 2d Session, 1950.

U. S. Temporary National Economic Committee, *Industrial Wage Rates, Labor Costs, and Price Policies,* Monograph No. 5, 1940.

U. S. W. P. A. National Research Project, *Mechanization, Employment and Output Per Man in Bituminous Coal Mining,* Report E-9, 2 vols., 1939.

———, *Trade Union Policy and Technological Change,* Report L-8, 1940.

United Mine Workers of America, *Nationalization of Coal Mines—The Miners' Program* [Washington?], Public Ownership League of America, 1922.

United Mine Workers' *Journal.*

United Mine Workers of America, *Proceedings* of the Constitutional Conventions.

U. S. News and World Report.

Van Kleeck, M., *Miners and Management,* New York, Russell Sage, 1934.

Virgin, R. Z., *Mine Management,* New York, Van Nostrand, 1922.

Wechsler, J. A., *Labor Baron. A Portrait of John L. Lewis,* New York, Morrow, 1944.

Wieck, E. A., *The Miners' Case and the Public Interest,* New York, Russell Sage, 1947.

INDEX